Geographic Information Systems (GIS) for Disaster Management

Geographic Information Systems (GIS) for Disaster Management

Brian Tomaszewski

CRC Press
Taylor & Francis Group
Boca Raton   London   New York

CRC Press is an imprint of the
Taylor & Francis Group, an **informa** business

CRC Press
Taylor & Francis Group
6000 Broken Sound Parkway NW, Suite 300
Boca Raton, FL 33487-2742

© 2015 by Taylor & Francis Group, LLC
CRC Press is an imprint of Taylor & Francis Group, an Informa business

No claim to original U.S. Government works

Printed on acid-free paper
Version Date: 20150128

International Standard Book Number-13: 978-1-4822-1168-9 (Hardback)

Library of Congress Cataloging-in-Publication Data

Tomaszewski, Brian.
　Geographic information systems (GIS) for disaster management / Brian Tomaszewski.
　　pages cm
　Includes bibliographical references and index.
　ISBN 978-1-4822-1168-9 (hardcover : alk. paper) 1. Emergency management--Geographic information systems. 2. Emergency management--Data processing. 3. Geographic information systems. 4. Disasters--Data processing. I. Title.

HV551.2.T647 2015
363.34'80285--dc23   2014039282

Visit the Taylor & Francis Web site at
http://www.taylorandfrancis.com

and the CRC Press Web site at
http://www.crcpress.com

CONTENTS

PREFACE

This book was primarily written for disaster management students interested in learning about the many facets of Geographic Information Systems (GIS) for disaster management. The unfortunate reality is that disasters will continue to proliferate in size, scope, and intensity. Future disasters will affect more people in diverse geographical contexts. Given that disasters are fundamentally spatial in nature, GIS plays a critical role in disaster management. However, there is an educational challenge and workforce need for well-educated practitioners and specialists who have a comprehensive, interdisciplinary understanding of the conceptual, technological, analytical, and representational capacities of GIS, as well as the policy and practice of disaster management. My hope is that this book can meet these challenges, even if partially. Additionally, I have made a particular point to gather a wide range of practical advice on developing a career in GIS for disaster management from experts ranging from local county government all the way to the United Nations. I strongly advise you to read their advice closely and keep it in mind as you develop and advance your own career if you are a student or use it to guide your students if you are a teacher. If you are GIS student interested in learning about the disaster management domain, the many examples provided in the book will ideally help you learn how GIS is applied to disaster management as well as more about GIS itself.

Furthermore, although the adoption of GIS into disaster management practice continues, there is still much more that can be done with integrating GIS and disaster management. Thus, beyond the book's primary audience, it has been designed to inform, enlighten, advocate for, and raise awareness of GIS for disaster management with working disaster management professionals, disaster management policy makers, and academic disaster management researchers with little to no understanding of GIS. GIS has the potential to advance interdisciplinary research and perspectives on disaster management due to the *spatial* nature of questions that GIS addresses and problems it helps to solve. Ultimately, it is my intent that anyone reading this book will develop better disaster management spatial thinking skills and learn how GIS can support spatial thinking. Specific GIS software titles will come and go, but it is the underlying spatial thinking skills for disaster management that will remain and are most important.

In 2003, when I was working as a GIS programmer in Buffalo, New York, I published my first short article on the topic of GIS for disaster management in a GIS trade magazine. At that time, I had no idea that 11 years later I would be writing the preface to a book-length treatment of the topic. It has been an incredible journey in the intervening years. There are many people to thank for helping me along the way to reach this point. I must give a note of gratitude to my doctoral advisor, Dr. Alan MacEachren of the Pennsylvania State University Geography Department and GeoVista Center, whom I will always consider my intellectual father and mentor for developing my abilities to produce a significant work of scholarship like this book. I must also give strong acknowledgment to my friend and colleague Lóránt Czárán, from the United Nations, without whom none of my fascinating and diverse United Nations research and other experiences over the past 7 years would have been possible. I thank the team at CRC Press, starting with Sarah Chow, who first contacted me about the book project idea. I would also like to give deep gratitude to Mark Listewnik from CRC Press for all of his extraordinary efforts in helping develop this book, especially when I faced a medical situation in 2013 that almost prevented the book's creation. I also thank Stephanie Morkert and Jennifer Abbott from CRC Press for helping see this book to final publication. I must also give great acknowledgment and gratitude to all of the book's interviewees for sharing their knowledge, experiences, and advice. Many of them spent many hours of their own time helping revise and edit their interviews, for which I am most grateful. It was a deep honor for me to have all of them participate in this project and I hope readers of this book will learn from their experiences. I also want to thank Dr. Anthony Vodacek and Dr. Jennifer Schneider from the Rochester Institute of Technology (RIT) for their help with reviewing book chapters.

I also thank my family and friends for all their support. Finally, I want to thank my soon-to-be wife, Allison Ramsay. Allison has been nothing but supportive and encouraging as I took on a massive book-writing project while still pretenure. She has endured the many long hours, often spent on weekends and evenings, with nothing but love and encouragement. By the time this book is finally published, we will be married and I look forward to a long and happy life with her.

This book is not a GIS software training manual. Rather, it is a book of ideas and examples that will show you what GIS is capable of doing for disaster management. Many good GIS software training books have already been written by the people and companies that create and sell GIS software. You are thus encouraged to find GIS software training books that match a particular GIS software title you're interested in (and perhaps learned about through this book) as a complement to ideas in this book. I have attempted to take advantage of the length that a book offers to provide a comprehensive treatment of GIS for disaster management. However, a single book cannot cover all aspects of this fascinating, interdisciplinary area. If there is something important that you think I missed, should discuss more, references that should be cited, or anything else, please contact me and tell me so; I would value your feedback. I hope that by reading this book you will learn as much as I did in writing it.

Brian Tomaszewski, PhD
Scottsville, New York, USA
bmtski@rit.edu

BIOGRAPHY

Brian Tomaszewski, PhD, is a geographic information scientist with research interests in the domains of geographic information science and technology, geographic visualization, spatial thinking, and disaster management. His published research on Geographic Information Systems (GIS) and disaster management–related topics has appeared in top scientific journals and conferences such as *Information Visualization, Computers, Environment and Urban Systems, Computers and Geosciences*, the *IEEE Conference on Visual Analytics Science and Technology,* and *The Cartographic Journal*. He also regularly publishes in popular GIS trade magazines such as *ArcUser* and *ArcNews*. He is also a scientific committee member for the Information Systems for Crisis Response and Management (ISCRAM) conference. His relevant experience includes past work with internationally focused organizations interested in GIS and disaster management such as the United Nations Office for the Coordination for Humanitarian Affairs (UN-OCHA) ReliefWeb service, United Nations Office for Outer Space Affairs Platform for Space-Based Information for Disaster Management and Emergency Response (UN-SPIDER), and United Nations Global Pulse. Dr. Tomaszewski also served as a visiting research scientist with the United Nations Institute for Environment and Human Security (UNU-EHS) in Bonn, Germany. He mentored and instructed multidisciplinary GIS for disaster management student research groups via the National Science Foundation (NSF)-funded Science Master's Program (SMP) titled Decision Support Technologies for Environmental Forecasting and Disaster Response at the Rochester Institute of Technology (RIT). His international research on socio-technical systems for displaced populations has been funded by the National Science Foundation (NSF), his research on geospatial technology educational development and spatial thinking in Rwanda has been supported by the United Kingdom Department for International Development (UK-DFID), and he is actively involved in other funded computing research activities in Rwanda. Dr. Tomaszewski is currently an assistant professor in the Department of Information Sciences and Technologies at the Rochester Institute of Technology. He holds a PhD in geography from the Pennsylvania State University. For more information, visit: http://people.rit.edu/bmtski/.

1

A Survey of GIS for Disaster Management

CHAPTER OBJECTIVES

Upon chapter completion, readers should be able to

1. understand the role of maps in disaster management,
2. describe how maps provide geographic context for disaster management and recognize how Geographic Information Systems (GIS) can be used for understanding geographic context,
3. be familiar with the concept of situation awareness,
4. discern the problems associated with the continued need for GIS in disaster management,
5. recognize the opportunities that exist with increased awareness and advocacy of GIS and mapping for disaster management, and
6. understand the importance of spatial thinking in disaster management practice.

INTRODUCTION

This book focuses on the application of GIS to disaster management. The book assumes no previous knowledge of GIS and little to no experience with disaster management ideas. To develop your skills and understanding of the application of GIS to disaster management, through the course of this book, you will learn about

1. scientific principles of geographic data and information,
2. how those principles apply to specific GIS software,
3. what GIS and related mapping software can and cannot do in terms of supporting disaster management practice,
4. how GIS relates to various disaster management cycle phases, and
5. ideas for keeping abreast of the ever-changing world of the application of GIS to disaster management.

1

In this chapter, a survey of GIS and disaster management is presented to get you thinking about some important concepts, followed by specific examples on the many ways in which GIS and mapping relate to disaster management.

GIS AND GEOGRAPHICAL CONTEXT

GISs have evolved into critical decision support and information management devices for all aspects of disaster management (National Research Council, 2007). This support and information management role comes primarily, although not exclusively, through the ability of a GIS to represent certain aspects of a disaster situation via maps. Maps in general have a long-standing role in disaster management—long before development of computerized GIS and digital data in general (Figure 1.1).

As in many domains such as engineering, urban planning, and the military, maps serve a fundamental purpose for understanding the *geographical context* of a disaster. The geographical context of a disaster can be thought of much like a news reporter asking for the basic *who, what, where, why,* and *how* aspects of a disaster situation (Tomaszewski and

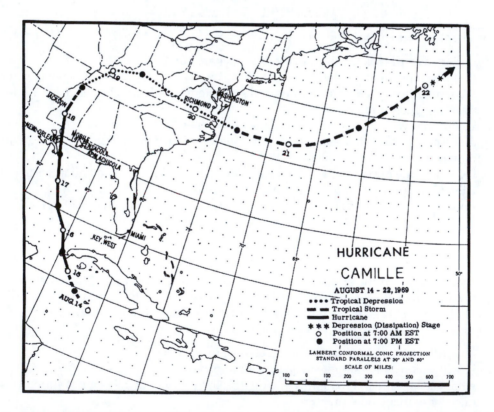

Figure 1.1 A 1969 tracking map of hurricane Camille. (From US Department of Commerce. 1969. *Hurricane Camille, August 14–22, 1969* [Preliminary Report], http://www.nhc.noaa.gov/archive/storm_wallets/atlantic/atl1969-prelim/camille/TCR-1969Camille.pdf [accessed May 22, 2014].)

MacEachren, 2012). First and foremost, maps can tell us the *where* aspect of a disaster—where are buildings damaged, where are roads open for evacuation, where are the areas that are most susceptible to flooding impacts, where should supplies be stationed for planning purposes. For many users of mapping tools in disaster management, the *where* aspect of maps is the most important function a map can serve. We will see many examples of the *where* aspect of mapping in the other parts of this book and, ideally, you will learn how to create basic maps that can show where things are happening in a disaster. However, it is important to consider that maps can also be used for deeper interpretation and reasoning of a disaster beyond simply showing where things are located.

For example, maps are also important for showing *what* is happening in a disaster and *when* it is happening. The *what* and *when* aspect of a map is particularly essential for showing *processes* during a disaster. As seen in Figure 1.1, a hurricane tracking map is used to show the weather categories (hurricane, tropical depression, storm) of the hurricane and how the hurricane will progress over time. This is a classic example of a map being used to show what is happening and when it is happening (as well where it is happening), and these types of maps are still very much used today. Other ideas about the *what* and *when* aspect of a maps in terms of disaster management activities to get you thinking beyond the *where* aspect of maps include

What is the extent of the flood?
What are the number of people impacted by the disaster?
What are the environmental processes at work in an area that are needed for mitigating the effects of a storm surge?
When will relief supplies arrive to the disaster area if they leave from a given distribution point?
What resources are available for disaster planning purposes?

The last two aspects of map use in disaster management are at a much deeper level and show how maps can facilitate disaster management decision making and reasoning. This involves the use of maps to understand the *how* and *why* about a disaster condition or situation. The following are representative examples of the *how* and *why* questions for disaster situations that maps can help answer:

How did an area become vulnerable to a disaster?
Why were the impacts from a disaster greater in one area compared to another?
How well will a disaster plan actually work in practice?
Why were there problems with the disaster response?
How can the physical environment best be mitigated against a natural hazard?
Why was recovery in one area slower than in other area?

Understanding the *how* and *why* about a disaster often involves a type of interaction process between the map reader and the map itself (MacEachren, 1995); for example, understanding and interpreting the symbols, colors, and other graphical aspects of a map (which are discussed in Chapter 2) to develop insight, reason, and make decisions. Modern GISs are key to this map user–map interaction process as GIS allows for dynamic interaction with a map and its data. For example, data layers can be quickly turned on and off or reordered for making comparisons to understand how a disaster evolved. Basic interactions such as panning and zooming allow areas of interest to be quickly viewed. Interactive querying capabilities allow for quick access to information that would otherwise be difficult to obtain. Map projections can be "projected on the fly" to incorporate and share data in varying formats with other disaster management teams. In many cases this can allow for greater understanding of a situation, swifter interpretation, and better—more informed—decision making. Statistical data displays can be quickly changed to reformat data and modify styling such as data class breaks and their color for reinterpretation of data (Figure 1.2).

In Figure 1.2, total counts of people aged 65–69 are shown in US counties. Such a map could be used for understanding where vulnerable populations, such as the elderly, are located for disaster planning purposes. Using GIS, the statistical display of the data class breaks can be quickly and easily manipulated. Note the top map, which assigns data observations to data class breaks based on equal numerical ranges (known as an *equal interval* classification), shows data outliers such as large population centers. The bottom map, which displays equal numbers of data observations per data class break (in this case, counties, and known as a *quantile classification*), gives a much different view of the data when compared to the equal interval map. Make particular note of the legends in each map and the differences between data observation assignments to data class breaks in each map. Having the ability to quickly modify the statistical display of data in GIS is

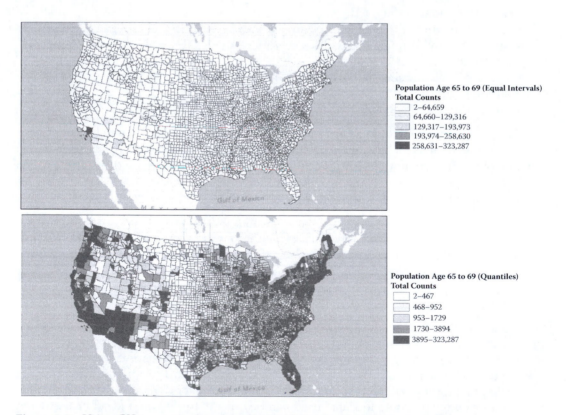

Figure 1.2 Using GIS to manipulate data display. (Maps by Brian Tomaszewski.)

one of the powerful aspects of GIS that will help you understand how and why a disaster situation developed. However, it also requires that a map maker and reader are aware of the effects of manipulating such displays, as seen in Figure 1.2, where exactly the same data can look very different depending on how it is classified and displayed. Maps such as these, also known as *thematic maps*, are discussed further in Chapter 2.

Using GIS to develop insight into *how* and *why* questions about disaster situations should be a long-term goal of anyone with a serious interest in using GIS for disaster management. As we will see in Chapters 3 and 8, more advanced use of GIS to answer *how* and *why* questions can be derived through the use of GIS models and understanding of the nature of geographic data and underlying geographic processes. *Situation awareness* is another concept that is closely related to the role of GIS for understanding a disaster's geographic context.

GIS and Situation Awareness

Today, maps can serve as the physical—and more often, virtual—representation of a disaster situation. The term *situation* can have multiple meanings. One perspective on the idea of a situation is that, in the context of GIS, a situation is the complete set of geographic, historic, and other factors that can potentially provide information and influence the actions of people working toward a goal (Brezillon, 1999).

A dictionary definition of *situation* is the "manner of being situated; location or position with reference to environment" (*Collins English Dictionary*, n.d.). Thus, *disaster situations* include *all* of the factors that must be accounted for by a disaster management team to guide and direct actions that are taken. For example, in a disaster response, the status of roads for relief supply delivery, the location of response teams and disaster victims, weather conditions, and the conditions of potentially damaged buildings. As one can imagine, the number of factors to be accounted for in a disaster (especially disaster response) could be endless. That is why GIS is an important device for supporting development of and providing *situation awareness* during a disaster, particularly when the information being used is updated in real time.

In its simplest form, situation awareness is knowing "what is going on." The term has strong military roots. In the military, understanding the situation—such as the position and status of troops, enemy locations, terrain, towns and infrastructure features such as roads and rivers, lines of battle, and other factors—are essential to decision making. The military has a long tradition of maps and cartographic conventions for displaying situations (Figure 1.3).

Figure 1.3 is a US military situation map from World War II, including the Allied landings at Normandy, France, in 1944. Make note of the following elements that represent the situation in this map: a clear date and time stamp as to when the map is displaying various elements, map symbols indicating the position of US and Allied units in relation to enemy units and the current front line of battle, and a small chart indicating "Units believed to be on the way to the battle area." Also, note how the current situation was hand drawn onto a base map of the current area of operation. Although an example from the military, this map demonstrates the value of situation awareness and how it could be of value in the context of disaster management and disaster response in particular.

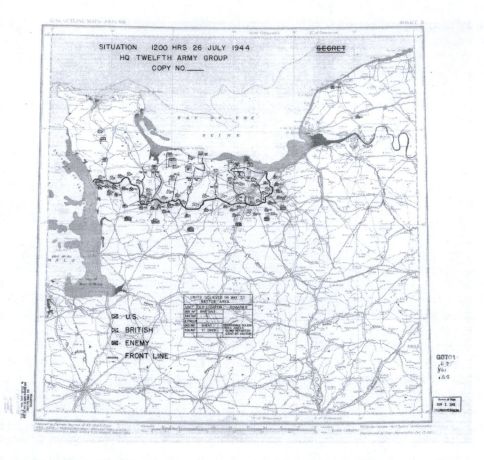

Figure 1.3 A historic US military situation map from World War II and the Allied landings at Normandy, France in 1944.

GIS can assist in the two stages of situation awareness. The first stage is *situation assessment*. Situation assessment is a process where information about the relevant factors in the environment is acquired. For example, GIS can be used to

- inventory initial damage assessments reported from field teams,
- acquire satellite and aerial imagery of a disaster zone,
- compile geotagged social media artifacts such as Twitter-based tweets and geotagged pictures, and
- organize news reports and citizens reporting.

Developing such assessments then leads to the second stage or that of actual *situation aware-ness*. Achieving situation awareness has been defined in the academic literature as the *compre-hension* of the state of the environment within a geographic extent (Endsley, 1995, 2000).

Disaster situations change over time. Thus, the process and interplay between situ-ation assessment and awareness is constant. GIS can play an essential role in managing the flow of information needed to help disaster management officials be aware of and assess situations. As discussed previously, the ability of GIS to quickly and easily manipulate and incorporate geographically referenced data is critical to disaster situation information management. For example, maps can be updated quickly to reflect changes in the situation such as the status of areas reviewed by disaster assessment teams. Imagery collected from manned or un-manned aerial or space-based remote-sensing platforms can be incorporated as it becomes available to aid in getting a *picture* of what is happening on the ground during a disaster response.

Increasingly in the US, disaster management officials are making the role and functions of GIS more accessible during disasters to provide real-time situation awareness—in disaster response in particular, but also during disaster planning and training exercises. In this regard, mobile GIS vehicles are a new and exciting development (Figures 1.4 a–d).

Operated by the GIS Services Division of Monroe County, New York, the vehicle depicted in Figures 1.4a–d is used to provide real-time situation awareness for emergency response and any other situations where county officials require real-time mapping, such as large-attendance public events. The vehicle has the following capabilities to support its mission. On the vehicle's exterior is a 30-foot mast with a pan/zoom/tilt camera, as well as a mast for a weather station. On the interior is storage and workspace for equipment such as workstation computers, tablet laptops, large-size (> 36 inches or 96 cm) printing (also known as a *plotter*), and large-screen LCD displays. The vehicle also has various office supplies, a microwave, and a refrigerator to support work staff. The vehicle can serve as a mobile wireless hotspot to support Internet access and can run off generators or connected power as available.

Despite exciting advances like the Monroe County GIS vehicle to support real-time situation awareness, there is a continued need for improvement of GIS in disaster management activities.

Figure 1.4a Monroe County, New York GIS technology vehicle. (Photo by Brian Tomaszewski. Used with permission from Monroe County GIS.)

Figure 1.4b View inside the Monroe County GIS technology vehicle looking toward the driver area. Note plotter on the left and work table on the right. (Photo by Brian Tomaszewski. Used with permission from Monroe County GIS.)

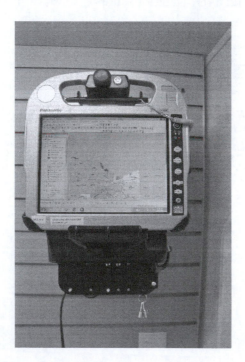

Figure 1.4c Close-up of field tablet computer docked inside the Monroe County GIS technology vehicle. (Photo by Brian Tomaszewski. Used with permission from Monroe County GIS.)

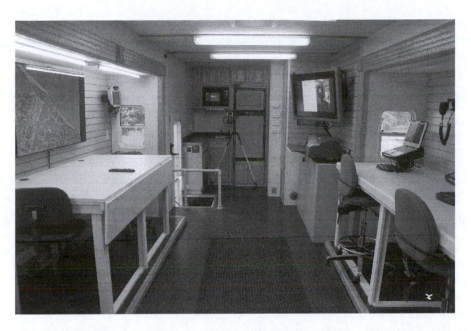

Figure 1.4d View inside the Monroe County GIS technology vehicle looking toward the back. Note work table on the left with a large mapping screen and GPS receiver base station in the rear. (Photo by Brian Tomaszewski. Used with permission from Monroe County GIS.)

THE PROBLEM: CONTINUED NEED FOR GIS IN DISASTER MANAGEMENT

Despite the well-documented benefits, there is still need for improvement in terms of further utilizing GIS for disaster management. In addition, the scope, scale, and intensity of disaster impacts continue to increase. Larger and increasingly diverse segments of society (as witnessed in the 2012 Hurricane Irene and Sandy impacts on the northeastern United States) are now being affected by disasters. These issues are even greater at the international scale and in the developing world. Improvements in the use of GIS for disaster management are needed in two primary areas: (1) general awareness of GIS technology in disaster management practice and the benefits it can provide and (2) improved coordination, sharing, and interoperability of GIS resources. The following sections discuss the ramifications of larger, more intense disasters; the need for improved coordination, sharing, and interoperability of GIS resources; and issues surrounding GIS awareness.

Scale, Scope, and Intensity of Disasters

Whether or not one believes that climate change is real, it is hard to argue against the fact that recent natural disasters such hurricanes and floods are becoming more intense and are affecting larger geographic areas and impacting larger segments of societies. This fact is now being recognized by government leaders, as evidenced by this 2012 quote from US New York State Governor Andrew Cuomo with regard to Hurricane Sandy (quoted in Vielkind, 2012):

> It's a longer conversation, but I think part of learning from this is the recognition that climate change is a reality, extreme weather is a reality, it is a reality that we are vulnerable. Climate change is a controversial subject, right? People will debate whether there is climate change … that's a whole political debate that I don't want to get into. I want to talk about the frequency of extreme weather situations, which is not political. … There's only so long you can say, 'this is once in a lifetime and it's not going to happen again.' The frequency is way up. It is not prudent to sit here, I believe, to sit here and say it's not going to happen again. Protecting this state from coastal flooding is a massive, massive undertaking. But it's a conversation I think is overdue.

Furthermore, the coupling of manmade and natural disasters, as evidenced by the 2011 Fukushima nuclear plant meltdown in Japan caused by an earthquake, revealed how vulnerabilities within critical infrastructures can compound natural hazard effects.

Changes in climate and weather conditions and their effects on natural hazards are even more pronounced at the international scale and in developing countries. Issues surrounding natural disasters in developing countries are made worse by the fact that many of these countries already have existing vulnerabilities and other issues such as political instability, famine, poverty, internally displaced persons (IDPs), refugees, and civil conflicts. The following case study highlights these issues.

Case Study: Burkina Faso—Disasters in the Developing World
Burkina Faso is a landlocked country in western Africa that is home to approximately 16 million residents (Figure 1.5).

As of 2009, it had a literacy rate of 26 percent, in 2010 an infant mortality rate of 91.7 per 1,000 live births, and an average life expectancy of 56.7 years. According to the US State Department, it is one of the poorest countries in the world. Burkina Faso is vulnerable to climatic shocks such as erratic seasonal weather patterns and longer-term global climate change that exacerbate natural disaster impacts. The primary natural disasters that Burkina faces are floods, drought, and locusts. The Conseil National de Secours d'Urgence et de Réhabilitation, Ministère de l'Action Sociale et de la Solidarité National (CONSAUR) is the national agency that deals with disaster damage assessments (victim identification, houses destroyed, etc.), humanitarian aid mobilization, natural disaster prevention and management training, socioeconomic infrastructure rehabilitation, and natural disaster victim needs assessments (PreventionWeb, 2013). CONSAUR is also active in developing a culture of disaster prevention, risk management, and societal resilience.

For example, Burkina Faso is developing a policy on National Multi-risk Preparation and Response to Disasters and a project to strengthen national disaster management capacities with the United Nations Development Programme (UNDP) (Conseil National de Secours d'Urgence et de Réhabilitation, 2010). Burkina Faso is also developing a disaster management project under the Global Facility for Disaster Risk Reduction (GFDRR) with the World Bank that is focused on developing insurance instruments to mitigate recurrent weather risk impact on small-scale cotton farmers (GFDRR, n.d.). Finally, Burkina Faso

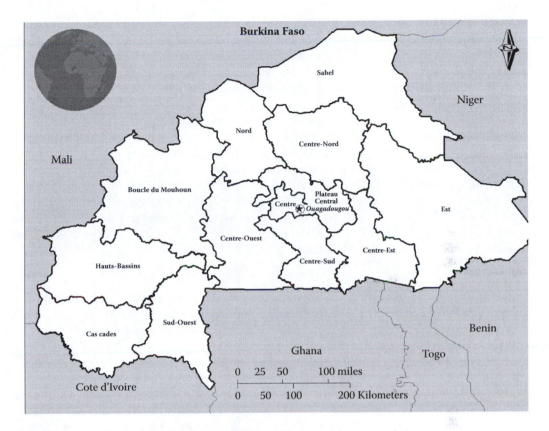

Figure 1.5 Overview map of Burkina Faso. (Map by Brian Tomaszewski.)

has adopted the Hyogo Framework for Action (HFA) to develop a general approach for reducing vulnerability to natural hazards (United Nations Office for Disaster Risk Reduction [UNISDR], n.d.).

Despite these and other efforts, national-, regional-, and local-scale disaster-coping capacities continually suffer from (CONSAUR, 2008, 2010):

- a lack of financial resources to fund local and regional initiatives,
- the late releasing of funds,
- a lack of material resources such as computers,
- insufficient support to decentralized structures (financial, material, logistics),
- insufficient qualifications of CONASUR, and
- decentralized structure staff.

A review of natural disasters and other types of crises in Burkina Faso since 2009 reveals the challenges associated with multiple, overlapping events from which it is difficult to recover. This is due to the general conditions in the country and the added challenges of lack of in-country technological capacity such as GIS to support relief and recovery efforts and a heavy reliance on outside assistance.

September 2009 Floods

In flooding that occurred in September of 2009, over 150,000 people were affected. Eight people were killed in the capital Ouagadougou. Water and electric systems were disrupted. The city's main hospital was partially flooded, and over 63,000 people sought temporary accommodation in schools, mosques, and churches (Office of US Foreign Disaster Assistance [OFDA], 2009) (Figure 1.6).

In terms of international GIS assistance and mapping support, the International Charter on Space and Major Disasters, which is a mechanism for providing satellite imagery of disaster zones to countries that do not have space-based assets (discussed further in Chapter 3), was activated to provide detailed maps of impacted areas (International Charter on Space and Major Disasters, 2013). Additionally, the MapAction group (http://www.mapaction.org, discussed further in Chapter 4), which is a nongovernmental organization (NGO) that deploys rapidly to disaster zones to provide mapping support, deployed to Burkina Faso and developed several detailed maps on flood inundation levels and other situational information (MapAction, 2011).

July and Early August 2010 Floods

In July and early August of 2010, heavy rains caused flooding in eight provinces of Burkina Faso killing 16 people and severely impacting 105,000 people (OFDA, 2010). In terms of external GIS and mapping support, the most publically available, detailed maps that were developed were produced by the German Center for Satellite-Based Crisis Information (ZKI or Zentrum für Satellitengestützte Kriseninformation, http://www.zki.dlr.de/mission), which is a service of the Remote Sensing Data Center (DFD or Deutsches Fernerkundungsdatenzentrum) of the German Aerospace Center (or DLR, Deutschen Zentrums für Luft-und Raumfahrt). These maps, produced through funding from the European Union, included flood damage extent maps in select cities and reference maps of the impacted area (Center for Satellite-Based Crisis Information [ZKI], 2010).

Figure 1.6 Flooding in Burkina Faso capital, Ouagadougou (2009). (Image © Brahima Ouedraogo/Integrated Regional Information Networks (IRIN), used with permission.)

Food Security (Ongoing)

Food security is a recurring problem in the broader Sahel region of West Africa where Burkina Faso is located and which is vulnerable to climate change effects (United Nations Environment Programme [UNEP], 2011). In 2011, increased food prices, low agricultural production, drought, and the inability of affected households to recover from the 2010 food price increases is making the region particularly vulnerable at the time of this writing (ReliefWeb, 2013). In terms of external GIS and mapping support, an excellent example of food security and famine mapping comes from the Famine Early Warning Systems Network (FEWS NET, http://www.fews.net). Operating with scientists in Africa, Central America, and the US, "the Famine Early Warning Systems Network (FEWS NET) is a United States Agency for International Development (USAID)-funded activity that collaborates with international, regional and national partners to provide timely and rigorous early warning and vulnerability information on emerging and evolving food security issues" (FEWS NET, n.d.). FEWS NET regularly publishes reports on issues related to food security such as food markets and trade, agro-climatic monitoring, livelihoods, and weather hazards. In terms of GIS and mapping, FEWS NET publishes maps on a quarterly basis related to food security outlooks, and they provide GIS datasets related to food security for download from their website (Figure 1.7).

Conflict in Mali (Late 2012 to Ongoing as of 2013)

At the time of the writing, Burkina Faso is facing a new crisis due to civil conflict in Mali, which borders Burkina Faso to the north and west. As of March 2013, over 48,000 refugees from Mali have crossed into Burkina Faso, creating strain on already stressed situations in Burkina Faso (Office for the Coordination of Humanitarian Affairs [OCHA], 2013). Groups such as the United Nations Office for the Coordination of Humanitarian Affairs (OCHA) have been active in developing maps of the Mali conflict and refugee situations.

Furthermore, many developing countries, like Burkina Faso, are significantly lacking in technological capacity to provide even basic information management capabilities such as GIS that can support disaster management activities. As will be discussed in the next section, significant issues still exist in the United States on geographic data sharing and coordination. These issues exist widely on the international scale but are even more challenging given the diversity of groups that are involved in providing support to developing countries during disasters.

According to the Emergency Events Database (EM-DAT), maintained by the Centre for Research on the Epidemiology of Disasters (CRED), worldwide and based on trend figures, the number of natural disasters reported has increased from approximately 75 in 1970 to approximately 400 in 2011. The number of people reportedly affected by natural disasters has increased from approximately 50 million in 1970 to approximately 300 million in 2011. Estimated damages caused by natural disasters and reported in US dollars have increased from approximately US$1 billion in 1970 to approximately US$350 billion in 2011 (Emergency Events Database [EM-DAT], 2009).

Thus, given the fact that natural disasters continue to increase, the dangers of coupled man-made and natural disasters, and increasing vulnerabilities in the developing world due to factors such as climate change, it is important to recognize the value and opportunity that GIS and mapping can play as a disaster management support mechanism in disaster management.

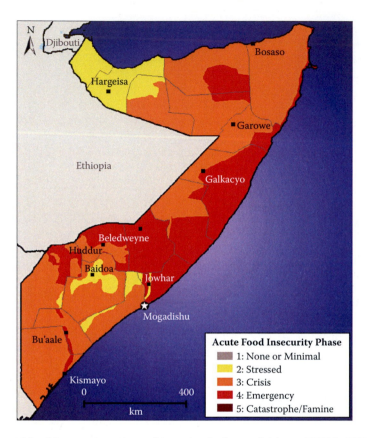

Figure 1.7 Map of food insecurity status of areas in the horn of Africa. FEWS NET generates maps like these using GIS and scientific models for predicting where food insecurity issues will arise. (From FEWS NET 2011.)

The Need for Improved Coordination, Sharing, and Interoperability

The greatest challenges exist in coordination, sharing, and interoperability of GIS resources such as data and trained staff and lack of comprehensive infrastructures for data sharing across local, state, federal resources (Cutter, 2003). These issues have been well documented and continue to be a primary issue hindering the use of GIS in disaster management.

The 9/11 terrorist attacks provide a key starting point for outlining specific coordination and sharing issues that still persist today. For example, damage caused during the 9/11 attacks to underground infrastructure such as pipes, water conduits, and subway tunnels had service ramifications for large parts of Manhattan and risk of electrocution, fire, and infrastructure collapse to rescue teams (Kevany, 2003). However, because information on underground infrastructure was not centrally maintained, much of the information was in formats totally incompatible with one another, in different vendor formats, and were owned by private organizations who were unwilling to share information for competitive reasons (Kevany, 2003). Similar issues arose during the response to Hurricane Katrina in 2005. During Katrina, disaster responders faced issues such as a lack of standardized and centralized information, a lack of data-sharing agreements in place, and a lack of common communication protocols (DeCapua, 2007).

On the international scale, the late 2004 Indian Ocean tsunami also revealed problems with data sharing and coordination. The massive scale of the Indian Ocean tsunami disaster, and the attention it received in the media, led to a massive outpouring of geographic data collection (remotely sensed imagery in particular). This mapping activity came from a worldwide spectrum of groups ranging from industry to academia, to NGOs, to international organizations such as the United Nations (UN), and government agencies from several countries (Kelmelis et al., 2006). Information and GIS data was not the problem. Despite these efforts, the sharing of data—and delivery of data and information to decision makers and first responders—was the primary problem. Many of the same issues as seen in 9/11 and Katrina arose during the response.

Kelmelis et al. (2006) outlined several additional issues. For example, some organizations were better equipped than others in terms of technical capacity to work with and analyze data that was collected. A common data repository and geographic products (i.e., maps) did not exist. Searching for data and products was time consuming. Lack of Internet access made base data difficult to obtain (if it existed at all) due to government restrictions. No common communications protocol existed, thus hindering communication across various groups such as the military, and civil government and societal organizations. Data collection standards were lacking as well as data quality due to lack of oversight of data review for field-collected data (Kelmelis et al., 2006). The Indian Ocean tsunami also revealed an issue that is becoming more prevalent in modern disasters—data quantity. The huge amounts of data made available by various data-generation groups often overwhelmed and created bottlenecks in various systems (Kelmelis et al., 2006). Data quantity issues were a key problem in the 2010 Haiti earthquake response and calls have been made in the international disaster management community for developing new techniques to triage the volume of data that is generated during a disaster so responders can focus on relevant tasks using relevant data and information (Harvard Humanitarian Initiative, 2011). In Chapter 6, techniques such as crowdsourcing are discussed as emerging ways in which large volumes of data such as aerial imagery collected in disaster zones can be used to analyze disaster situations.

The 2007 National Research Council report *Successful Response Starts with a Map: Improving Geospatial Support for Disaster Management* also made several key recommendations from improving data coordination and sharing (National Research Council, 2007). The following is a summary of those recommendations (the parenthetical information at the end of each item refers to the recommendation number in the original report):

1. the US Department of Homeland Security (DHS) taking an more active leadership role via a National Spatial Data Infrastructure (NSDI, discussed in Chapter 4) framework to develop policies and procedures to ensure that a wide spectrum of agencies have access to geographic data and tools for all phases of disaster management (recommendations 1, 2, and 3);
2. development of security procedures to ensure that data such as critical infrastructure is shared with appropriate stakeholders (recommendation 4);

3. development of standing contracts and procurement procedures across local, state, and federal scales for acquisition of disaster event data such as remotely sensed imagery and other geographic data and information (recommendation 5); and

4. emphasis on intensive preparedness exercises across all groups involved in disaster management to address cultural, institutional, procedural, and technical problems associated with communication across groups and the complexity of geographic data challenges during a disaster (recommendation 6).

Problems of GIS Awareness in Disaster Management

There is room for improvement in regard to making disaster management practitioners—and ultimately the broader public—aware of the power and benefits of GIS and mapping in general. The benefits of GIS have been well documented in the academic literature. The 9/11 terrorist attacks demonstrated the very useful benefits of GIS, but also *it limits rescue, response, and recovery efforts in an extreme situation* (Kevany, 2003). The 9/11 attacks led to advances in research on how three-dimensional (3D) GIS can be used to navigate within multilevel structures to help emergency responders search buildings (Kwan and Lee, 2005).

Hurricane Katrina in 2005 pointed out different types of GIS awareness issues. When Katrina struck, responders suffered the effects of not having plans for incorporating GIS into the response. For example, base maps were almost 10 years out of date; there was initially a complete lack of GIS funding; and GIS professionals were not incorporated into the response, with a heavy reliance being made on GIS volunteers (DeCapua, 2007). Katrina also pointed out problems of GIS awareness in terms of the culture of disaster management, such as lack of technical training and technology aversion—issues identified by other disaster management researchers (Cutter, 2003). The following quote from the geospatial lead in Baton Rouge during Katrina summarizes the culture issue (quoted in DeCapua, 2007, 37):

> One problem is that there are cultural differences between old and new school views. People who work in mitigation don't see the use of the GIS tools available. Technology isn't embraced. Mitigation and preparedness is generally done by local and state so FEMA cannot enforce it.

Since Katrina, there has been a steady increase in the awareness of GIS for disaster management and outside of academic literature and GIS for disaster management career opportunities continue to grow. The aforementioned 2007 National Research Council (NRC) report, *Successful Response Starts with a Map*, was a key development for GIS awareness raising. Developed with input from academia, industry, NGOs, and government officials, the key conclusions of the report were that GIS and related technology and tools should be an essential part of all aspects of disaster management, however, (1) lack of preparation for future events and (2) immediate needs such as saving lives, shelter, and food often take precedence over mapping, and thus training, coordination, and resource investments often lack priority by decision makers (National Research Council, 2007). In terms of awareness of GIS in disaster management practice within the United States, the NRC report made several key recommendations that point to an optimal vision of how GIS might be used;

these recommendations are summarized as follows (the parenthetical information at the end of each item refers to the recommendation number in the original report):

1. GIS should be formally included in disaster management agency planning policies and procedures (recommendation 1);
2. academic organizations that provide emergency management curricula should make a greater emphasis on GIS technology (recommendation 9);
3. the US Federal Emergency Management Agency (FEMA) should expand and retain full-time GIS staff who can quickly deploy to help respond to event (recommendation 10);
4. DHS should maintain a secure inventory of qualified GIS professionals who can support disaster response activities (recommendation 11), and
5. federal funding and grants should be increased to support state and local governments for GIS preparedness activities (recommendation 12).

In the following section, opportunities for GIS and mapping in the disaster management context are discussed based on examples of the broader increased awareness and advocacy of GIS and mapping in general.

THE OPPORTUNITY: INCREASED AWARENESS AND ADVOCACY OF GIS AND MAPPING

In 2005, 77 percent of emergency operation centers (EOCs) at the state level in the United States had one or more staff members assigned to GIS applications (Hodgson, Davis, and Kotelenska, 2010; cited in Hodgson et al., 2013). A 2011 survey conducted by the Department of Geography/GIScience Research Laboratory at the University of South Carolina indicated that at the state level, all EOCs were utilizing GIS and remote sensing to varying degrees due to increased funding, awareness, and coordination and changes in technology (Hodgson et al., 2013).

To highlight some specific examples of how GIS continues to be further integrated into the activities of disaster management practitioners—and how recognition of the benefits of GIS continues to increase—FEMA now offers an online course (titled IS-922: Applications of GIS for Emergency Management) that provides a general overview of GIS and emergency management (Federal Emergency Management Agency [FEMA], 2012). In terms of disaster management policy in the United States, GIS and geospatial data are now explicitly referred to in several official policies such as the National Incident Management System (NIMS), the National Response Framework (NRF), and others. The formal role of GIS within disaster management policies within the United States and international contexts is further discussed in Chapters 4 through 8. Chapter 9 provides practical advice from experts on building a career and finding a job in GIS in the disaster management field.

Outside of the disaster management practitioner community, GIS and mapping in general are seeing a growing trend in use by people typically not trained in traditional mapping science disciplines such as geography. For example, academic researchers from disciplines such as information technology, computer science, and political science, NGOs and the general public are continuing to embrace the power of maps and spatiel thinking.

This trend is closely linked to recent changes in mapping technology. These changes are allowing mapping capabilities to be available to a wider range of people than traditional GIS software (discussed in Chapter 3), which often takes months, if not years of training to become proficient with, and in the case of commercial GIS software, can be restrictive in terms of procurement costs. Technology such as Google Maps are lowering the barriers for creating and utilizing digital maps. Now, anyone can make a map. This is a good development and yet also demands caution as it is easy now for anyone to make a bad map due to ignorance of cartographic design, science, and geographic data representation principles. For example, tools such as Google Maps Engine (https://mapsengine.google.com) allows a user, for free and without any need for computer programming, to add points, lines, polygons, pictures, and hyperlinks to the Google Maps base map and share the map with anyone. Map makers with some knowledge of computer programming languages such as JavaScript can build custom applications to be run on the web or mobile devices using the Google Maps application programming interface (API) (https://developers.google.com/maps/). Increasingly, free, web-based mapping tools such as Google Maps are being used by those who are referred to as "neographers" (or new geographers looking beyond traditional GIS approaches) to create mapping "mashups" (or the combination of myriad data sources onto a map) to develop a variety of mapping approaches such as space–time maps that integrate social media and public participation and feedback (Liu and Palen, 2010). Chapter 3 further discusses the ideas of mapping mashups and technology such as the Google Maps API and other mapping APIs. One particular recent develop in the awareness of mapping outside traditional disaster management communities is *crisis mapping*.

CRISIS MAPPING

Although a specific origination date is unknown, the notion of *crisis mapping* is believed to have begun with development of the Ushahidi (which means "testimony" in Swahili) mapping platform during the postelection violence in Kenya in late 2007 and early 2008. Due to government bans on media and self-censorship in the mainstream media, an information vacuum soon emerged in regard to ethnic violence that was occurring after the elections (Okolloh, 2009). Thus, Ushahidi was developed to facilitate a map-centric approach to the crowdsourcing of information about reports of violence. Ushahidi allowed people (or the "crowd") to make reports about events happening to a central website using Short Message Service (SMS) technology or through interacting with the Ushahidi website directly (Okolloh, 2009). As reports came in, they were approved by Ushahidi site administrators to remove any false or erroneous reports, and then the reports were displayed on a map with events symbolized based on the event type (Okolloh, 2009).

Since this initial beginning, the ideas of crowdsourcing and crisis mapping have expanded. Crisis maps are now commonplace for major disasters. As disasters and other crises around the world continue to escalate, online, crowdsourced mapping continues to proliferate—and the efforts are even beginning to attract the attention of the mainstream media (Lohr, 2011). Additionally, the crisis mapping approach continues to play an important role in international crisis situations where a lack of on-the-ground media coverage or restrictive government control of information creates information gaps for understanding what is actually happening. Often, volunteers from around the world (who are not necessarily GIS experts) work at mapping events into crisis maps to help develop a broader picture of a crisis situation. For example, a crisis mapping volunteer will monitor media reports and social media (i.e., Facebook and Twitter) for any information that could be relevant to incorporate into a crisis map. Recent examples of the power of the crowdsourcing/crisis map approach for filling information gaps are the 2010 Haiti earthquake (Zook et al., 2010), the 2011 Libyan civil war and at the time of this writing, the civil war conflict in Syria (Figures 1.8a and 1.8b).

Figures 1.8a and 1.8b are 2013 crisis maps from the civil war conflict in Syria, also known as the Syria Tracker. Figure 1.8a is an overview of the overall Syria Crisis Map. Make note of the following in this image: an overview map indicating the number of reports received per area using a clustering technique, a graph below the map indicating the frequency of reporting made by day, and on the right, report categories. Clicking one of the report categories will filter the map display to show only reports matching the selected category. Figure 1.8b is a detail of the map shown in Figure 1.8a. In this detail, the map has zoomed in on Damascus, a major city in Syria, and the map has been filtered to show reports of killings. An individual report has been clicked, indicating that 42 people were killed in the vicinity of the black circle icon shown on the map. Note that the map

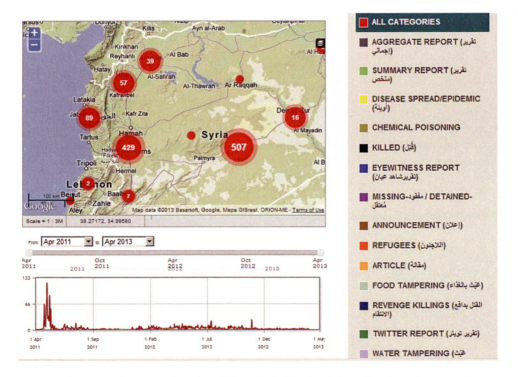

Figure 1.8a Syria Tracker overview map. (From Syria Tracker, https://syriatracker.crowdmap.com; a Project of Humanitarian Tracker, http://www.humanitariantracker.org; used with permission.)

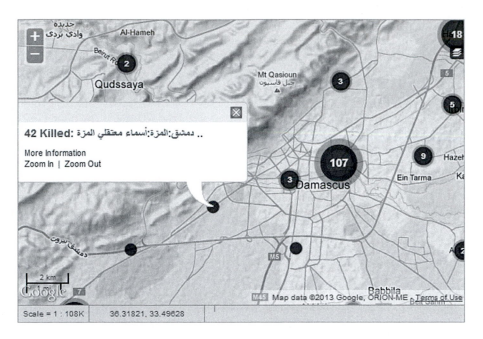

Figure 1.8b Syria Tracker detail map. (From Syria Tracker, https://syriatracker.crowdmap.com; a Project of Humanitarian Tracker, http://www.humanitariantracker.org; used with permission.)

clustering technique for indicating the number of points in a given area is also active when the map is zoomed in. This is a very useful feature for managing the display and interaction with multiple points that share the same coordinate, a common cartography problem (Tomaszewski, 2009). The overall Syria Tracker Crisis Map is a key information-gathering platform for the situation in Syria due to dangerous conditions on the ground for outside media and government restrictions or false reporting.

The following interview from a leading crisis mapping researcher and thought leader provides further perspectives on crisis mapping.

Interview with Dr. Jennifer Ziemke, Cofounder and Codirector of the International Network of Crisis Mappers

Dr. Jennifer Ziemke (Figure 1.9) is a leading scholar in the field of crisis mapping. Her research applies spatial and temporal econometric analysis, dynamic visualization, and in-depth historical and archival research to develop maps that reveal underlying complex processes. Dr. Ziemke served as a Peace Corps volunteer on the Namibian side of the Angolan border from 1997 to 1999 and has extensive experience in a dozen African countries.

She is cofounder and codirector of the International Network of Crisis Mappers, co-organizer of the International Conference on Crisis Mapping (ICCM) series, and an assistant professor of international relations at John Carroll University (JCU). She is also a Fellow at the Harvard Humanitarian Initiative (HHI) and consults for a number of international organizations in the United States and Europe.

Figure 1.9 Jennifer Ziemke PhD, Political Science, University of Wisconsin-Madison.

The following is the first of a two-part interview conducted for this book with Dr. Ziemke in April 2013. In this portion of the interview, she answers questions about the broader impacts of crisis mapping on the raising of awareness of maps and mapping of disasters for wider audiences, opportunities for further incorporation of GIS in crisis mapping, the impacts of crisis mapping on disaster management practice and the work of disaster management professionals, and challenges and issues associated with crisis mapping. The second part of this interview will be presented in Chapter 9 where Dr. Ziemke discusses the near- and long-term future of crisis mapping.

Crisis mapping (CM) has been gaining attention across diverse groups of people outside traditional disaster management practice; how do you think CM has helped raise awareness of the power of maps/mapping to wider audiences?

University and high school students tend to be captive and engaged audiences. As youth are constantly connected to social media on mobile devices, the idea that we can learn something from gathering and mapping this data easily resonates with them. So, taking that extra step and saying "OK, so how would visualizing this information on a map help an organization respond to a disaster, or monitor an election, or bear witness to other kinds of events?" These questions make sense to students.

Students I have met at both my university and beyond are enthusiastic to help in any way they can—whether with language translation, data georeferencing, or cleaning data in a micro-tasking environment. After teaching them the basics of media monitoring, the second point of discussion always turns toward the importance of GIS. Basic crisis maps that display many red dots are better visualized and analyzed inside a GIS, so we show students how to import data into a GIS platform for further analysis.

In general, many different groups around the world are learning about the power of crowdsourced mapping, particularly in crisis environments. From media outlets like Al Jazeera to local women's groups and affected populations, many have begun to embark on a project of monitoring that often includes maps.

What opportunities do you see for further integration of "traditional" GIS practice, such as the work done by GIS professionals and/or GIS educators, and CM?

From the beginning, GIS experts have been an important part of our network and community. The Crisis Mappers Network was launched at the end of our first annual conference in 2009, [the International Conference of Crisis Mappers, or ICCM]. We continue to dialogue and learn from people whose primary expertise is in GIS. Members include those working with both the open-source and proprietary variants of the software. While many of our members are highly skilled in this practice and the value of this practice is recognized by the community, more can be done in terms of seamless integration—where everyone, including new members and young volunteers, comes to understand the important role that GIS and GIS education plays in visualizing and analyzing this data. Universities are a great place where better integration between these worlds can happen. The demand for courses on crowdsourcing, crisis mapping, and humanitarian response is growing, as are the complimentary demands for good courses on GIS, cartography, data analysis, and remote sensing.

Do you think CM has been having any impact on traditional disaster management practice and the work of disaster management professionals? If so, how?

Since 2009, the Crisis Mappers network has engaged disaster management practitioners from a wide range of backgrounds, across different areas of expertise, regional deployments, and organizational perspectives. Summarizing developments across thousands of different institutional environments, and in a global and extremely diverse community, is nearly impossible. But I do see a number of new conversations being raised as a result of shared concerns on the Crisis Mappers Google Group, and am always hearing about a bewildering proliferation of new projects and maps.

I think that disaster managers have different perspectives on all of the core debates, including data verification, privacy, security, liability, and effective service provision. One of the things that Crisis Mappers Net provides is a global venue or forum for conversations and debates about these topics. As a result, best practices and lessons learned are being shared. For example, just today we hosted a webinar in which the head of the Protection of Civilian Populations Unit at the International Committee of the Red Cross (ICRC) in Geneva shared the results of their advisory group's review on topics as diverse as informed consent, data interpretation, data manipulation, and the assessment of risk. Some institutions are leaders at the forefront of engaging these debates and propelling them forward, while still other groups eagerly respond with their own lessons learned and additional concerns.

Although showing promise, what issues/problems/challenges (if any) do you see with any aspect of CM?

In an ideal world, we would get the news out to everyone that yes, you can make your maps and collect live data to populate these maps as part of your organization's overall strategy, but first you need to think carefully, and ask yourself, What is the purpose of the map? What is this project about? What is the best strategy?

Maybe your project takes place in an extremely sensitive environment, like a war zone, where your tolerance for inaccuracy is low, or the security and privacy concerns for respondents are too high. In this case, actively encouraging the crowd to submit sensitive data might not be the best strategy. Crisis maps don't necessarily need to rely on crowdsourced data alone. One could train and use a trusted network, for example, instead.

While problems around privacy, security and data verification are nearly always present, in most cases I think they can be mitigated with careful planning. Just because a concern arises about privacy, liability, or security does not mean you should not visualize these data on a map. Rather, it means to take seriously the efforts by many in this community to consider each of their well-defined steps and incorporate them into a carefully designed project in a way that sufficiently minimizes the risk. And on the question of data verification, even if your tolerance for inaccuracy for your project is extremely low, you have to ask yourself, is it better than nothing at all? Maybe some of my map contains rumors or false information. But even if some of the data are false, perhaps we can still learn a great deal from the overall trends. These questions depend on the tolerance you have for risk, and that level will vary from project to project. A crowdsourced map of the best burgers in Cleveland may contain some misleading information, but we can tolerate risk in this case.

After carefully considering the purpose and goals of your project, you may decide that a geographic map is actually not the best visualization of the data pouring in about the live event. In the case of the recent attack at the Boston Marathon, for example, a map was quickly deemed by the global, virtual community to be less useful per se than other forms of crowdsourced sharing and offers of assistance. Instead, various organizations jumped in and offered help. Google Person Finder was deployed to help connect missing persons, whereas other volunteers stood up and shared spreadsheets to collect offers of assistance and free places to stay. The crowd was sharing, and volunteer Crisis Mappers were engaged in everything from posting public announcements on Twitter to helping with emotional trauma, but all of this activity took place in Skype chat rooms, on cell phones, and inside spreadsheets, and not on a map.

In the past few years, there has been an "exhilaration" around the idea that "we can map everything in real time" without carefully thinking through the purpose first. As the case with the Boston Marathon attack shows, we need to be careful not to waste time and resources creating maps where our energy would be better spent in a different direction. In sum, we need to be very deliberate about the purpose of the proposed map. Next, we need to try to find out whether other communities have already stood up a map, so as to avoid duplication of effort, and work with the other digital volunteers who have already mobilized around the event, using Twitter, Google groups, and YouTube, and within minutes. These volunteers are diffuse, global, often highly skilled, but they should be engaged.

SPATIAL THINKING AND DISASTER MANAGEMENT

The use of GIS for disaster management can ultimately be seen as a means for developing spatial thinking skills. Spatial thinking is the idea of using the properties of space (distance, direction), visual representations (maps, diagrams), and reasoning processes to structure and solve problems (National Research Council, 2006; Bednarz and Bednarz, 2008). We use spatial thinking every day—whether we are aware of it or not. For example, counting using one's fingers, planning road trips and reasoning about travel distances using a map, or using space as a metaphor for time (i.e., "the event is far off in the future") (National Research Council, 2006).

The need to "think spatially" is perhaps the most important skill anyone interested in the application of GIS for disaster management can learn. Disasters are inherently spatial problems—the interactions between people, the places they live, the environments that surround them, the events that affect them, and the broader networks of support that are drawn upon, are all multiscale space–time phenomena that must be considered in any aspect of disaster management. Examples of spatial thinking and the relationship with disaster management include making judgments about the safest and shortest routes for evacuation planning, reasoning and developing understanding of the spatial relation-ships between natural hazards and how they can potentially impact infrastructures in built environments, and thinking about abstract spatial relationships such as the inter-connections between the physical environment and development and planning (Berse, Bendimerad, and Asami, 2011).

GIS plays an important role for development of visual representations and interactions with geographic data and information that can facilitate spatial thinking processes. As we have seen through many examples in this chapter, GIS and mapping in general are a start-ing point for enabling spatial thinking about disaster problems—whether it is using GIS to quickly change the display of geographic statistical data to understand the distribution of vulnerable populations (Figure 1.2), understanding the positions and movements of troops in a battle (Figure 1.3), developing measures of food insecurity in Africa for informing relief and recovery efforts (Figure 1.7), or gathering information on the civil conflict in Syria to make the world aware of what specifically is happening and where (Figures 1.8a and 1.8b). Ultimately, the use of GIS for disaster management should be seen as a means to an end and not the end itself—the end being the development of spatial thinking skills, supported with GIS, to understand and reason about geographic-scale relationships to solve problems and make effective decisions.

CHAPTER SUMMARY

In this chapter, you were given a survey of GIS for disaster management. You first learned that GIS is critical to understanding the geographical context of disaster situations. You also learned the important role GIS has in answering who, what, where, why, and how questions related to disaster management. You are encouraged to keep these types of questions in mind as you read through subsequent chapters and learn more about the various ways in which GIS can be used for disaster management.

You also learned about an idea that will be repeated throughout the book—situation awareness. GIS is vital to establishing and maintaining knowledge of what has happened during a disaster situation, often in the form of maps. You were then given some perspec-tives on various problems that still persist in GIS for disaster management. For example, larger and more intense disasters of international scope, challenges with coordination and sharing of geographic information, and the continued need for further advocacy of the capabilities and benefits of GIS. More importantly, you then learned about the many great opportunities that exist for GIS for disaster management. In the past 10 years, GIS continued to prove its value for disaster management, and the future is bright for interesting jobs and career opportunities that exist at the intersection of GIS and disaster management, a topic you will learn extensively about throughout this book. In this regard, you saw an introduction to the exciting topic of crisis mapping, which involves using new forms of mapping technology to gather large volumes of geographic information from volunteers, or "the crowd." You then read an interview with one of the leading thinkers on crisis mapping to get you thinking about the possibilities that crisis mapping offers to disaster management.

The chapter ended with another important idea you will hear again throughout this book—spatial thinking. Always remember that GIS tools are support devices to enable and support your spatial thinking and reasoning to make better-informed decisions and judgments. The next chapter discusses the fundamentals of geographic information and mapping. It is important that you have an understanding of the core scientific principles that underlie GIS technology. This will enable you to adapt to changes in specific GIS software over time as ideas and scientific principles of geographic information, such as coordinate systems, map projections, and cartography have existed for centuries and will continue to be the foundation of GIS software.

DISCUSSION QUESTIONS

1. What are some other specific examples of who, what, where, when, why, and how disaster management geographic context questions that GIS could help answer?
2. What challenges are there with maintaining situation awareness, and how specifically can GIS support situation awareness?
3. Explain some of the issues associated with GIS support for international disaster management.
4. What are some of the main coordination and sharing issues with GIS in disaster management practice?
5. Why are there problems with awareness of GIS in disaster management practice?
6. Describe some of the opportunities that are emerging for GIS and disaster management.
7. Explain the importance of spatial thinking to disaster management practice.

REFERENCES

Bednarz, Robert S., and Sarah W. Bednarz. 2008. "The importance of spatial thinking in an uncertain world. In *Geospatial Technologies and Homeland Security*," Springer.

Berse, Kristoffer B., Fouad Bendimerad, and Yasushi Asami. 2011. "Beyond geo-spatial technologies: Promoting spatial thinking through local disaster risk management planning." *Procedia-Social and Behavioral Sciences* 21:73–82.

Brezillon, Patrick. 1999. "Context in artificial intelligence: II. Key elements of contexts " *Computer & Artificial Intelligence* 18 (5):425–446.

Center for Satellite Based Crisis Information (ZKI). 2010. "Flood in Burkina Faso." Center for Satellite Based Crisis Information, July 13, http://www.zki.dlr.de/article/1524 (accessed March 31, 2013).

Collins English Dictionary. "Situation." HarperCollins Publishers, n.d., *Collins English Dictionary* online, http://dictionary.reference.com/browse/situation (accessed March 27, 2013).

Conseil National de Secours d'Urgence et de Réhabilitation, Ministère de l'Action Sociale et de la Solidarité National (CONASUR). 2008. "Rapport d'Activites 2008 du Secretariat Permanent du Conseil National de Secours d'Urgence et de Rehabilitation" (Unpublished report).

Conseil National de Secours d'Urgence et de Réhabilitation, Ministère de l'Action Sociale et de la Solidarité National (CONASUR). 2010. "Rapport d'Activites 2010 du Secretariat Permanent du Conseil National de Secours d'Urgence et de Rehabilitation (Sp/CONASUR)" (Unpublished report).

Cutter, Susan L. 2003. "GI science, disasters, and emergency management." *Transactions in GIS* 7 (4):439–446.

DeCapua, Chelsea. 2007. *Applications of Geospatial Technology in International Disasters and during Hurricane Katrina.* Oak Ridge National Laboratory.

Emergency Events Database (EM-DAT). 2009. "Natural disasters trends," Emergency Events Database, http://www.emdat.be/natural-disasters-trends (April 1, 2013).

Endsley, Mica R. 1995. "Toward a theory of situation awareness in dynamic systems." *Human Factors* 37 (1):32–64.

Endsley, Mica R. 2000. "Theoretical underpinnings of situation awareness: A critical review." *Situation Awareness Analysis and Measurement*:3–32.

Federal Emergency Management Agency (FEMA). 2012. *IS-922: Applications of GIS for Emergency Management.* FEMA, http://training.fema.gov/EMIWeb/IS/courseOverview.aspx?code=is-922 (accessed March 28, 2013).

FEWS NET. 2011. "SOMALIA food security outlook April to September 2011," edited by United States Agency for International Development (USAID), http://www.fews.net/sites/default/files/documents/reports/Somalia_OL_2011_04.pdf (accessed May 22, 2014).

FEWS NET. n.d. "What is FEWS NET?" FEWS NET, http://www.fews.net/ml/en/info/Pages/default.aspx?l=en (accessed April 1, 2013).

Global Facility for Disaster Reduction and Recovery (GFDRR). n.d. *GFDRR Case Study: Burkina Faso,* GFDRR, https://www.gfdrr.org/docs/Snapshots_Burkina_Faso.pdf (accessed May 22, 2014).

Harvard Humanitarian Initiative. 2011. *Disaster Relief 2.0: The Future of Information Sharing in Humanitarian Emergencies.* Washington, D.C. and Berkshire, UK.

Hodgson, Michael E., Sarah E. Battersby, Shufan Liu, and Leanne Sulewski. 2013. *Geospatial and Remote Sensing Data Use by States and Counties in Disaster Response and Recovery: A Nationwide Survey,* http://people.cas.sc.edu/hodgsonm/Published_Articles_PDF/Survey%20Geospatial%20Remote%20Sensing%20Data%20Use%20State%20Counties_2-20-2013.pdf (accessed May 22, 2014).

Hodgson, Michael E., Bruce A. Davis, and Jitka Kotelenska. 2010. "Remote sensing and GIS data/information in the emergency response/recovery phase." In *Geospatial Techniques in Urban Hazard and Disaster Analysis,* Dordrecht: Springer.

International Charter on Space and Major Disasters. 2009. "Floods in Burkina Faso," International Charter on Space and Major Disasters, http://www.disasterscharter.org/web/charter/activation_details?p_r_p_1415474252_assetId=ACT-265 (accessed March 31, 2013).

Kelmelis, John A., Lee Schwartz, Carol Christian, Melba Crawford, and Dennis King. 2006. "Use of geographic information in response to the Sumatra-Andaman response to the Sumatra-Andaman earthquake and Indian Ocean earthquake and Indian Ocean tsunami of December 26, 2004." *Photogrammetric Engineering & Remote Sensing* 72 (8):862–877.

Kevany, Michael J. 2003. "GIS in the World Trade Center attack: Trial by fire." *Computers, Environment and Urban Systems* 27 (6):571–583.

Kwan, Mei-Po, and Jiyeong Lee. 2005. "Emergency response after 9/11: The potential of real-time 3D GIS for quick emergency response in micro-spatial environments." *Computers, Environment and Urban Systems* 29 (2):93–113.

Liu, Sophia and Leysia Palen. 2010. "The new cartographers: Crisis map mashups and the emergence of neogeographic practice." *Cartography and Geographic Information Science* 37 (1):69–90.

Lohr, Steve. 2011. "Online mapping shows potential to transform relief efforts." *New York Times*, March 28, http://www.nytimes.com/2011/03/28/business/28map.html?_r=0 (accessed April 2, 2013).

MacEachren, Alan M. 1995. *How Maps Work: Representation, Visualization, and Design*, New York Guilford Publications.

MapAction. 2011. "Burkina Faso flooding, September 2009," MapAction, http://www.mapaction.org/deployments/depldetail/186.html (accessed March 31, 2013).

National Research Council. 2006. "GIS as a support system for spatial thinking." In *Learning to Think Spatially: GIS as a Support System in the K–12 Cirriculum.* Washington, D.C.: National Academies Press.

National Research Council. 2007. *Successful Response Starts with a Map: Improving Geospatial Support for Disaster Management.* Washington, D.C.: National Academies Press.

Office for the Coordination of Humanitarian Affairs (OCHA). 2013. "MALI: Humanitarian snapshot (as of 25 March 2013)," ReliefWeb, March 25, 2013, http://reliefweb.int/report/mali/mali-humanitarian-snapshot-25-march-2013 (accessed May 22, 2014).

Office of US. Foreign Disaster Assistance (OFDA). 2009. *Annual Report for Fiscal Year 2009.* ReliefWeb, http://reliefweb.int/report/world/annual-report-fiscal-year-2009-office-us-foreign-disater-assistance-ofda (accessed May 22, 2014).

Office of US Foreign Disaster Assistance (OFDA). 2010. *Annual Report for Fiscal Year 2010.* United States Agency for International Development, http://pdf.usaid.gov/pdf_docs/pdacs473.pdf (accessed May 22, 2014).

Okolloh, Ory. 2009. "Ushahidi, or 'testimony': Web 2.0 tools for crowdsourcing crisis information." *Participatory Learning and Action* 59 (1):65–70.

PreventionWeb. 2013. "Burkina Faso National Platform," PreventionWeb, http://www.prevention-web.net/english/hyogo/national/list/v.php?id=27 (accessed April 1, 2013).

ReliefWeb. 2013. "Sahel: Food insecurity 2011–2013," ReliefWeb, http://reliefweb.int/disaster/ot-2011-000205-ner (March 31, 2013).

Tomaszewski, Brain, and Alan MacEachren. 2012. "Geovisual analytics to support crisis management: Information foraging for geo-historical context." *Information Visualization* 11 (4):339–359.

Tomaszewski, Brian. 2009. "Managing multiple point-based data instances on a single coordinate in mapping mashups." Paper read at the 24th International Cartography Conference (ICC), at Santiago, Chile.

United Nations Environment Programme (UNEP). 2011. *Livelihood Security: Climate Change, Conflict and Migration in the Sahel.* Geneva, Switzerland: United Nations Environment Programme.

United Nations Office for Disaster Risk Reduction (UNISDR). n.d. "Hyogo Framework for Action (HFA)," UNISDR, http://www.unisdr.org/we/coordinate/hfa (accessed April 1, 2013).

US Department of Commerce. 1969. *Hurricane Camille, August 14–22, 1969* (Preliminary Report), http://www.nhc.noaa.gov/archive/storm_wallets/atlantic/atl1969-prelim/camille/TCR-1969Camille.pdf (accessed May 22, 2014).

Vielkind, Jimmy. 2012. "Cuomo: 'Climate change is a reality … we are vulnerable.'" 2012. Capitol Confidential, October 31, http://blog.timesunion.com/capitol/archives/162798/cuomo-climate-change-is-a-reality-we-are-vulnerable/ (accessed March 29, 2013).

Zook, Matthew, Mark Graham, Taylor Shelton, and Sean Gorman. 2010. "Volunteered geographic information and crowdsourcing disaster relief: A case study of the Haitian earthquake." *World Medical & Health Policy* 2 (2):7.

2

Fundamentals of Geographic Information and Maps

CHAPTER OBJECTIVES

Upon chapter completion, readers should be able to

1. understand the difference between data and information;
2. describe the concept of map scale and how map scale is represented;
3. understand what map projections are and discern the differences between different map projection types;
4. be familiar with common map coordinate systems;
5. understand mapping principles such as data measurement, visual variables, and figure/ground relationships;
6. discern the differences between reference and thematic maps; and
7. identify common errors when first learning to use Geographic Information Systems (GIS) to create maps.

INTRODUCTION

This chapter presents important scientific principles related to the fundamentals of geographic information and maps. For centuries, geographers and others have had a core need to represent human, natural, and other activities at the earth's surface. This need has led to the development of several core concepts and ideas that guide representation of the earth and can have strong influence on how those representations (i.e., maps) affect interpretation of mapped items. For example, and as you will see later in this chapter, the choice of map projection can have significant consequences on how a given piece of geography is represented in terms of shape and size.

The concepts and ideas you will learn about in this chapter, all of which existed long before the advent of computers and GIS, are the foundations of how the earth is represented in any format—whether it is a two-dimensional (2D) paper map or a complex three-dimensional

(3D) virtual environment. As you will see in later chapters, the concepts and ideas presented in this chapter will continually resurface in both general, software-agnostic concepts for how GIS is applied to disaster management, but also in specific GIS software products. Thus, developing a good understanding of these fundamentals will (1) improve your abilities in applying GIS to disaster management practice, (2) help you understand the limits and relevancy of geographic data you may encounter, (3) help you to develop your vocabulary for common terms used in GIS and mapping applications, and (4) ultimately, make you a better spatial thinker.

The chapter structure is as follows. First is a discussion of map scale, or how a map takes ground measurements and proportions them into map units. Next is a discussion of map projections, or how the earth (which is a 3D circular entity) is represented in a flat, 2D map format, and the issues associated with this transformation. Coordinate systems are then discussed to demonstrate how specific locations are spatially indexed and referenced. Based on these core scientific principles of geographic information, the chapter then provides an overview of mapping and cartography fundamentals. Reference maps are discussed, which are general purpose maps (also known as *base maps*) and thematic maps, which present one or more variables or interests or convey a "message." Finally, basic cartographic principles are discussed given the importance of making usable maps that do not miscommunicate.

Data vs. Information

Before proceeding, it is important to distinguish between and define two terms used often in upcoming chapters—data and information. Although often used interchangeably, data and information are not the same. *Data* are raw facts or observations. Examples of data include weather temperatures or rain fall volume. *Information* is data with context or making sense of data so that it is actionable or useful. Using the proceeding examples, information derived from weather temperature data would be a weather report. Maps can be thought of as a form of information artifact as maps contextualize data into visual formats. Data are fundamental to GIS. If you are new to GIS or a seasoned GIS professional, you will most likely spend a majority of your time with GIS software working with handling, creating, or editing data. GIS, however, works with both data and information. For example, the underlying resources that are incorporated into GIS (such as road locations or building footprints) are data. These data are then contextualized into information in the form of a digital map. In Chapter 3, further discussion is made on how GIS transforms data into information. For this chapter, the term *information* is used to describe the fundamental principles of earth representation as many of these principles derive from data, for example, individual latitude and longitude points that comprise a map projection or the *x,y* coordinates that form a coordinate system. The following section discusses these and other important geographic information principles.

Scale

Map scale is a ratio (or proportion) between measurements on the map and corresponding measurements on the ground. Thus, the idea of map scale is exactly the same concept of scale used in the hobby of model building or any other domain where a real-life entity is represented in a reduced (or modeled) manner.

Three Ways of Representing Map Scale

There are three common ways of representing map scale in both hard-copy and digital mapping environments.

1. *Representative fraction:* The representative fraction is the relationship between map and ground units and thus does not represent any specific map units, such as feet, meters, miles, or kilometers. When map scale is represented as a representative fraction or ratio, the scale is presented numerically such as 1:100,000 or 1/1,000,000. A representative fraction such as 1:100,000 is interpreted as "one map unit is equal to one hundred thousand ground units." When applying specific units to a representative fraction, the same units must be used for both the map and ground numbers. Using the previous representative fraction example of 1:100,000, if you wanted to know how many inches in ground units are represented by 1 inch on the map, you would interpret the representative fraction scale as 1 inch of map units is equal to 100,000 inches on the ground.

2. *Verbal statement:* Expressing map scale as a verbal statement is simply the idea of stating the map scale in a verbose manner such as "one inch to sixteen miles." Verbal statement scales are of limited use for calculating ground distances in units other than the units given in the verbal statement.

3. *Graphic bar:* Perhaps the most common and useful form of expressing scale is with a graphical bar or other visual device (Figure 2.1).

Graphic bars allow for a map reader to visually inspect distances displayed on the graphic bar and quickly compare those distances with map distances to determine ground distance, thus eliminating the need for mathematical calculations. Another important

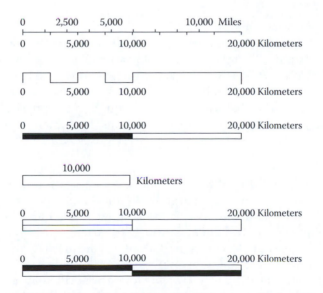

Figure 2.1 A nonexhaustive collection of graphical scale bars used in both digital and hard-copy maps.

CALCULATING GROUND DISTANCE ON MAP USING THE MAP SCALE

Calculating a land distance on a paper map is easy to do. As a starting point, the following formula is useful:

$$Sm = \frac{Dm}{Dg}$$

where
    Sm = The map scale
    Dm = Map distance
    Dg – Ground distance
    Remember that Dm and Dg are the same measurement units

Using the above formula, see if you can solve the following problem:

Imagine you are a walker, and you want to determine how far you walked today. You measure the walking path on your map and the path is 150 millimeters long. Your map scale is 1:24,000. What is the length of the real walking path?

Hint: Plug the numbers given into the formula and calculate the representative fraction:

$$\frac{1}{24,000} = \frac{150}{?}$$

Answer: You walked 3,600,000 mm or 3.6 km. The ground distance was calculated by multiplying 24,000 * 150

aspect of graphic bar scales is that they can adjust as the extent of the map changes. For example, if an 11-inch (30 cm) by 17-inch (43 cm) paper map is physically reduced in size, the graphic bar scale will be reduced accordingly and still be relevant for calculating ground distance from the map using the graphic bar scale. The same cannot be said for representative fraction and verbal scale representations because physical changes to a map alter the proportions in these types of representations. In digital mapping, graphic scale bars change as the zoom level of the map changes (Figures 2.2a and 2.2b).

Large- vs. Small-Scale Maps
As mentioned at the beginning of this chapter, it is important to develop a vocabulary around core mapping terminology. Large-scale and small-scale maps are important terms you should understand. Small-scale maps show a larger area with less detail. For example, a map of the entire world at a scale of 1:30,000,000 printed on a 8.5-inch (22 cm) by 11-inch (30 cm) page would be considered a small-scale map. A large-scale map shows a smaller area with more detail. For example, a 1:24,000-scale map like those found in the United States Geological Survey (USGS) topographical map series. The ideas of small- and large-scale maps might seem counterintuitive at first. One might think that a map showing a larger area would be classified as a large-scale map and a smaller area

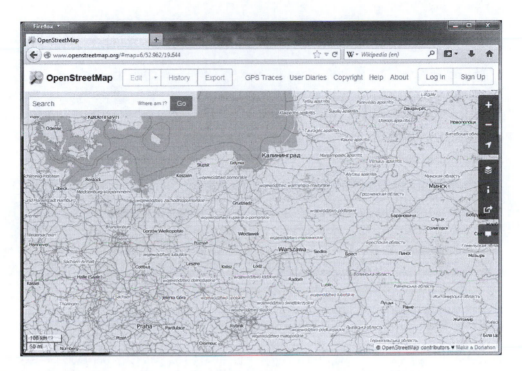

Figure 2.2a In this online map, the graphical scale bar indicates the distance of map units to ground units. (From © OpenStreetMap contributors.)

a small-scale map. One way to think of the distinction is that items on a small-scale map are less detailed or smaller—much like the features on the ground look smaller when looking out the window when flying in an airplane. The closer one is to a ground feature, the larger it appears. Figure 2.3 visually illustrates the differences between small-scale and large-scale maps.

Why Scale Matters: Detail and Accuracy

For disaster management applications, there are two reasons why scale is important. The first is that *data detail is scale dependent*. Looking back at Figure 2.3, the scale, and in turn the details of each map, change the relevancy of each map for different purposes. Using a disaster management example, the 1:250,000-scale map on the far left of Figure 2.3 might be effective for planning relief supply transportation routes on major highways, the 1:100,000-scale map in the middle would be effective for planning where to station relief supplies around a village, and the 1:24,000-scale map on the right would be effective for evacuation planning of specific locations such as houses or neighborhoods within a village. Each of these tasks is dependent on the map scale used to accomplish the task. Furthermore, as we will see in Chapter 3, when the concept of metadata (or data that describes a dataset) is discussed, many digital datasets used in GIS are based on the data digitization from hard-copy sources. Thus, it is important to know the scale at which the data were digitized at to ensure that the data detail is sufficient for the purpose for which the data is being used.

Figure 2.2b In this view, the map from Figure 2.2a has been zoomed in but the web browser window containing this map has not changed size. Graphical scale bars that dynamically change are now commonplace in many digital mapping applications and are very useful for quickly getting a sense of the extent of geography being shown on a map. (From © OpenStreetMap contributors.)

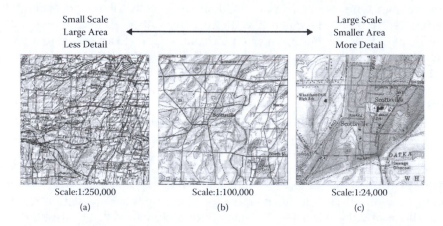

Figure 2.3 The differences between small- and large-scale maps. Note how the map on the left shows a larger overall area but with less detail. As the map scale changes (by moving to the right through the figure), small areas are shown but with more detail. These maps are taken from the USGS topographic map series. (a) 1:250,000 is taken from http://gis.ny.gov/gisdata/quads/drg250/c43076a1. htm. (b) 1:100,000 is taken from http://gis.ny.gov/gisdata/quads/drg100/f43077a1.htm. (c) 1:24,000 is from http://gis.ny.gov/gisdata/quads/drg24/p17.htm. (accessed September 15, 2014.)

For example, a road network digitized from a 1:250,000-scale paper map will not have the same details as a road network digitized from a 1:24,000-scale map.

The second issue is that *accuracy is scale dependent*. For example, if using a paper map or a digital source that is based on a paper map to locate building positions for disaster planning, it is vital that the map is dependable in terms of the building locations shown on the map actually being near where they located if their locations were field verified using precise ground surveying. In fact, the USGS has long published map standards to ensure vertical and horizontal accuracy of map points. More specifically, "the horizontal accuracy standard requires that the positions of 90 percent of all points tested must be accurate within 1/50th of an inch (0.05 centimeters) on the map. At 1:24,000 scale, 1/50th of an inch is 40 feet (12.2 meters). The vertical accuracy standard requires that the elevation of 90 percent of all points tested must be correct within half of the contour interval. On a map with a contour interval of 10 feet, the map must correctly show 90 percent of all points tested within 5 feet (1.5 meters) of the actual elevation" (US Geological Survey, 2006).

MAP PROJECTIONS

The earth is an oblique sphere. Thus, starting with the ancient Greeks and the beginnings of earth measurement and graphical representation, a practical need developed for map projections. A map projection is a method of representing the Earth's three-dimensional surface as a flat two-dimensional surface.

The following list outlines key points you should know about map projections:

1. Projections are mathematical transformations (will be discussed later in this chapter).
2. Scale is true only in certain places.
3. Many different types of projections have been devised.
4. All map projections distort.
5. Distortion characteristics vary among projection types.
6. Some types are better for some applications than others.
7. A few types are used widely.

The idea of "projecting" a map is that a (hypothetical) light source is placed in the center of a 3D scale model of the Earth. From this center point, the light would then shine out (or project) the lines of latitude and longitude onto a 2D projection surface, which would then be the basis for the map projection. All map projections are based on three common projection surfaces, which are illustrated in Figures 2.4a–2.4c:

In Figures 2.4.a–2.4c, the heavy black lines (or in 2.4c, the black point on the top of globe) are known as *standard lines*. Standard lines are where the map projection surface touches the globe model. The standard lines are the only place on the map where the scale is true (Figure 2.5).

Creating a map projection distorts the spatial properties of the final 2D map. Spatial properties that are distorted include distance, area, shape, and direction. When a flat map is made, choices are made, depending on the intended use of the map, as to which of these properties to preserve. Thus, distortion exists in all flat 2D maps as no map projection can preserve all these properties. Map projections are often classified based on

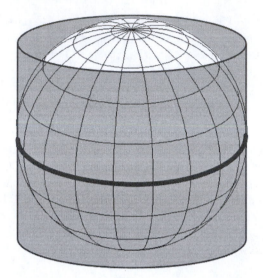

Figure 2.4a   Cylinder projection surface, used for cylindrical projections. In this case, the projection surface is tangent, or in contact with a single point or line of the globe.

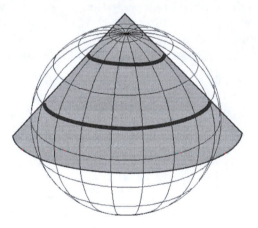

Figure 2.4b   Cone projection surface, used for conic projections. In this case, the projection surface is secant with the globe, or cutting/intersecting through the globe.

the spatial property they preserve. The following is a list of map projection categories (Environmental Systems Research Institute, 2010; National Atlas of the United States, 2013; Carlos A. Furuti, 2008; Robinson et al., 1995):

Conformal: This projection preserves shape and angles. Conformal projections are useful for maps such navigation charts where preserving the shapes of small areas and angles is important. Shapes of large areas are distorted.

Equal area (equivalent): This projection preserves area and size. Equal area projections are useful for maps that show large areas of land, such as continents.

Figure 2.4c Plane projection surface, used for planer or azimuthal projections. In this case, the projection surface is tangent with the globe.

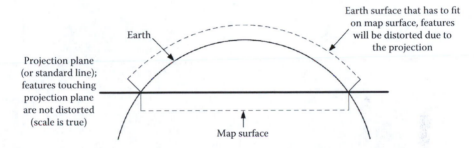

Figure 2.5 The relationship between the projection plane, Earth surface, and map surface. (Adapted from Environmental Systems Research Institute. 2010. What is a map projection? ArcGIS Desktop 9.3 Help, http://webhelp.esri.com/arcgisdesktop/9.3/index.cfm?topicname=what_is_a_map_projection? [accessed April 2, 2014].)

Equidistant: This projection preserves true distances in some directions from the projection center or along special lines. As an example, an azimuthal equidistant map centered on New York City will show the correct distance to Washington, D.C. or any other point. The map would also show the correct distances from New York City to Buffalo, but would not show the correct distance between Buffalo and Washington, D.C.

Azimuthal: This projection preserves true direction (azimuths or angle measurements) from a reference point. Azimuthal can be combined with other projection properties, as seen in the previously mentioned azimuthal equidistant projection.

Another category of map projection is the *compromise projection*. Compromise projections try to strike a balance in distorting the various spatial properties.

Figure 2.6 graphically illustrates map projections and the trade-offs they make in distortions of various spatial properties.

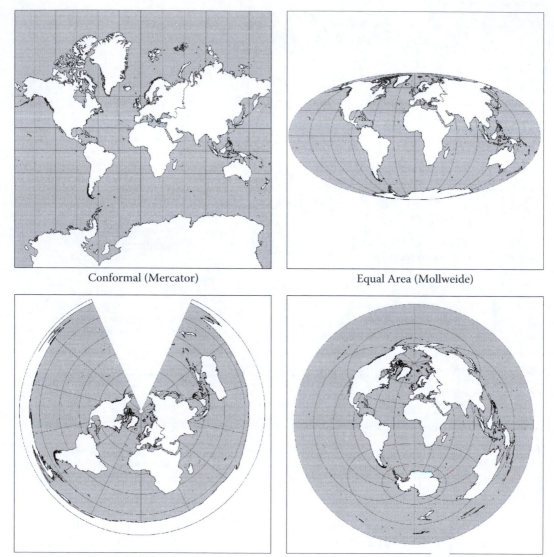

Conformal (Mercator)

Equal Area (Mollweide)

Equidistant (Equidistant Conic)

Azimuthal (Azimuthal Equidistant)

Figure 2.6 Map projection examples. In these figures, the general projection class is listed with the name of the specific projection shown in parentheses. The conformal map projection (top left), clearly illustrates how shapes are preserved but size and area are greatly distorted, as seen in the size and area of Greenland and Antarctica when compared with Africa. The equal area projection (top right) shows features at their correct size and area, but not their correct shape. This time Africa is shown correctly in terms of size and area when compared with Greenland and Antarctica. The equidistant projection (bottom left) clearly demonstrates the use of conical projection surface. The azimuthal projection (bottom right) is centered on 0° latitude and 0° longitude, and thus any direction measured from this point will be correct. (All maps by Brian Tomaszewski.)

At this point, you may be wondering which projections are most widely used. The following are general guidelines that you can follow when having to make a decision as to which map projection to choose:

- Use true angle (conformal) projections for large-scale maps and planar coordinate systems (discussed in the next section).
- Use true area (equal area) and compromise projections for small-scale maps of the Unites States and the world.
- Use true distance (equidistant) and true direction (azimuthal) projections for special projects such as navigation.

COORDINATE SYSTEMS

Throughout history, maps have served a fundamental purpose for geographically referencing and indexing locations at the surface of the earth. Geographical referencing at the surface of the earth has many forms, from zip codes, to street addresses, to latitude and longitude coordinates. Latitude and longitude coordinates are particularly important as they are the basis for many other types of geographical referencing such as 2D map projections and subsequent coordinate systems based on 2D map projections (discussed in the next section). To develop a map projection, latitude and longitude coordinates are mathematically converted into geographic planar Cartesian (*x,y*) coordinates.

Latitude and longitude coordinates, also referred to as *spherical coordinates*, use the measures of angles (degrees) from both the center of the earth for latitude and from the prime meridian (or zero degrees) for longitude. Figure 2.7 graphically demonstrates the idea of latitude and longitude coordinates.

Planar coordinates are based on the ideas of the Cartesian space and referenced on an *x,y* grid. In a Cartesian *x,y* grid, the *x*-axis are east-to-west coordinate values, and the *y*-axis is north-to-south coordinate values. Figure 2.8 graphically demonstrates the basic idea of a Cartesian *x,y* grid, which is the basis for planar coordinates. Later in this chapter, you will see specific examples of planar coordinate systems.

You may be wondering why planar coordinate systems were developed, since they are based on map projections which, as we have seen, inherently contain errors. The reason planar coordinates were developed is that they are more efficient and provide better meaning for measurement than spherical coordinates. For example, in an application such as surveying, it is more meaningful to express distances and areas in terms of feet or meters versus degrees, minutes, and seconds. In the following section, two common planar coordinate systems are discussed.

Universal Transverse Mercator Coordinate System

The Universal Transverse Mercator (UTM) coordinate system is an international standard planar coordinate system. In the UTM system, the Earth is divided into 60 zones that span 6° of longitude each (60 zones * 6° = 360° total covering the entire Earth) and each UTM zone is divided into a north and south section (Figure 2.9).

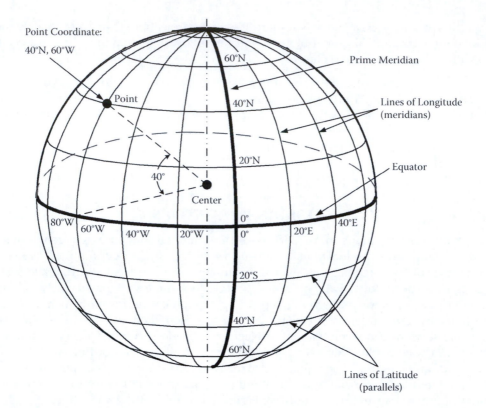

Figure 2.7 The latitude and longitude coordinate system. Lines of latitude (also known as parallels) are measured from north to south based on their position in relation to the equator. Lines of longitude (also known as meridians) are measured east to west from the prime meridian. In this figure, a hypothetical point is being referenced at 40°N, 60°W. Note how this point is derived based on angular measurements—latitude from the center of the Earth and longitude as degrees from the arbitrary starting point of the prime meridian (0°). (Adapted from National Oceanographic Partnership Program [NOPP]. n.d. Track a NOPP Drifter, http://galileospendulum.org/2011/page/41/. [accessed April 2, 2014].)

Each zone uses its own transverse cylindrical projection (meaning the cylindrical projection plane is turned to be around the North and South Poles and not the equator) to minimize scale distortions (refer back to Figure 2.5). Figure 2.10 graphically demonstrates the characteristics of a single UTM zone.

Although the UTM coordinate system is very useful due to its ability to internationally reference any point on the earth via a planar coordinate system, it also has drawbacks. The most notable drawback is that the 60 zones of the UTM system do not conform to political boundaries or jurisdictions, and can thus be unusable in situations where a geographical referencing system is needed across an entire political or jurisdictional entity. Furthermore, given that each UTM zone is defined by its own unique projection, maps of adjoining zones will not conform to one another along a shared border. Figure 2.11 demonstrates this issue for UTM zones that span the continental United States.

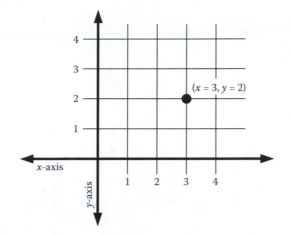

Figure 2.8 The Cartesian grid. In this example, a point is referenced at an *x* value of 3 and a *y* value of 2. Note that this (*x,y*) coordinate is in the positive number space of the grid. Planar coordinate systems are based on the ideas of a Cartesian grid.

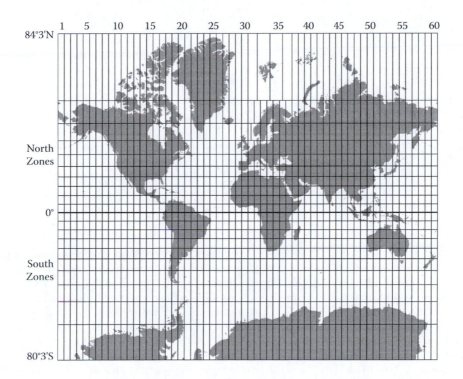

Figure 2.9 The UTM World Zone grid shown on a world Mercator projection. The zones are labeled in increments of 5 at the top of this figure. Note how the zones become distorted the closer they are to the north and south poles. The equator (0° latitude) is used to mark the boundary between the north and south sections of each UTM zone. (Map by Brian Tomaszewski.)

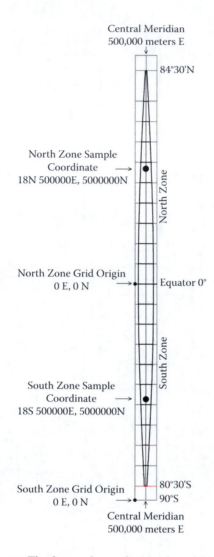

Figure 2.10 A single UTM zone. The figure shows the North and South sections of the zone, which are divided by the equator. Note how the Cartesian (*x,y*) grid of the North zone originates on the bottom left of the zone on the equator (0° latitude) and the Cartesian (*x,y*) grid of the South zone originates on the bottom left of the zone at 90° south. From the respective (0,0) origin points of each zone, coordinates are then measured out in meters along the zone's projection in a positive number space (as seen in Figure 2.8) and in units referred to as *northings*, or measurements from north to south, and *eastings*, or measurements from east to west. The central meridian of each UTM zone is referred to as a *false easting* (so all the coordinate values will be positive) and is assigned the value of 500,000 meters E. Finally, make note of the two sample UTM coordinates shown in Figure 2.10 that are approximately in the middle of the North and South, respectively. These sample coordinates demonstrate the how a UTM coordinate pair is written in terms of indicating (1) the zone number, (2) the north or south hemisphere (N or S), (3) the six-digit easting (E) value, and (4) the seven-digit northing (N) value. For example, the North Zone Sample Coordinate is 18N 500000E, 5000000N.

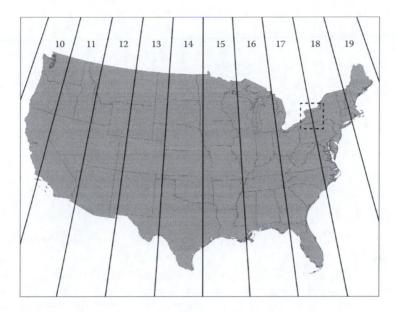

Figure 2.11 UTM zones in the continental United States. Note how many of the zones divide various states into two or more sections. For example, make note of the dashed box shown on the upper right of the figure that highlights the area between zones 17 and 18, which cross through New York State. In this case, a special map projection called *UTM Zone 18 extended* has been developed so that UTM zone 18 coordinates can be used for all of New York State. (Map by Brian Tomaszewski.)

State Plane Coordinate (SPC) System

Like UTM, the State Plane Coordinate (SPC) system is based on a series of specialized map projections that define specialized zones. In the SPC, however, all of the zones are within the United States and defined within political boundaries (Figure 2.12).

As you may recall from the discussion on projection surfaces, distortion is least along the standard lines. Thus, state plane zones use specialized projections optimized to fit the shape and orientation of the zone contained within that state (Figure 2.13).

Datums

The final important concept that you need to understand for coordinate systems are datums. A *horizontal datum* (which would derive coordinates in the *x,y* plane) consists of two elements—a reference ellipsoid and accurately known control points.

Reference Ellipsoids
The earth is not a perfect sphere. It is shaped more like an ellipse (egg), and therefore its shape must be approximated in order for coordinate systems to be referenced to the earth's surface. *Reference ellipsoids* are mathematical approximations of the earth's shape; many have been developed over the past 200 years and are often given names of the mathematician that developed the ellipsoid, such as the Clarke 1866 reference ellipsoid.

Figure 2.12   State Plane Coordinate (SPCs) zones within the United States. Note that zones do not extend beyond state boundaries and that a state may have several zones. Thus, SPCs are not suitable for regional (i.e., multiple-state) mapping. State plane zones generally measure coordinates in US feet values. (Map by Brian Tomaszewski.)

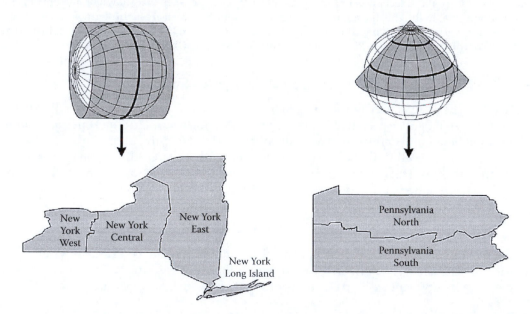

Figure 2.13   Examples of SPCs and the projection surfaces used to define those zones. The left side of this figure shows SPCs for New York. These zones are based on a transverse cylindrical projection as they are north-to-south oriented zones. The right side of this figure shows SPCs for Pennsylvania. These zones are based on a cylindrical projection surface as they are east-to-west oriented zones. The choice of a particular projection surface is made to minimize scale distortion caused by the projection. (Maps by Brian Tomaszewski.)

Technical discussion of reference ellipsoids is beyond the scope of this book (see the topic Geodesy in this chapter's Resources section). It is important to understand the basic idea of ellipsoids in a GIS context because a spot on the earth's surface can have different coordinate values based on the reference ellipsoid used to measure the coordinates. Figure 2.14 graphically demonstrates the idea of reference ellipsoids.

Control Points

Control points are accurately measured locations used as reference points in land surveying and for developing datums. In the United States, government organizations like the US Coast and Geodetic Survey physically mark control points with a small metal disk called a *benchmark* (Figure 2.15).

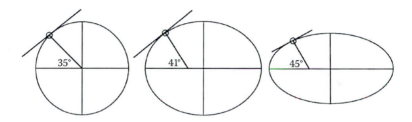

Figure 2.14   Reference ellipsoid examples and their importance within a coordinate system context. Note how the example point (in the figure above, the small circle) located at the same spot on the surface has different latitude values (35°, 41°, and 45°, respectively) based on the reference ellipsoid used to measure it.

Figure 2.15   Example of a benchmark.

The Importance of Datums

Datums, which you will often encounter in a GIS context, have been developed based on advances in accurate earth shape measure and to cover wide areas. Common datums you will find in GIS datasets include the North American Datum of 1983 (or NAD 83) and the World Geodetic System 1984 datum (or WGS 84). WGS 84 is the datum that is used for most GPS receiver coordinates. It is very important to know what the datum is when working with GIS data, as you saw in Figure 2.14. Different datums based on different reference ellipsoids can cause the same location to have significantly different coordinate values depending on the coordinates that the datum references (Figure 2.16).

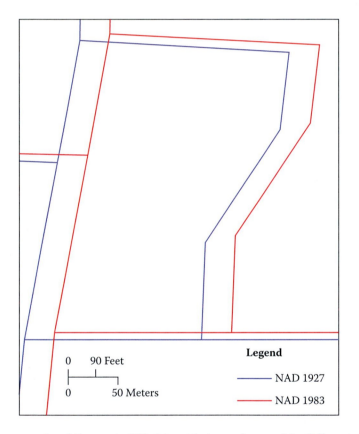

Figure 2.16 An example of the same GIS dataset being referenced in different datums and the issues this can cause. In this figure, the blue lines are a road network referenced in the North American Datum of 1927 (NAD 27) and the red lines are the same road network referenced in the North American Datum of 1983 (NAD 83). Note the how the NAD 83 data appear almost "shifted" from the NAD 27 data. Locations can have an almost 100-foot (30-meter) difference in where their coordinates are referenced depending on the datum used. Checking what the datum is for a GIS dataset is a good first step for troubleshooting data that does not overlay properly. Not knowing the datum is a common problem that beginning GIS users encounter when working with GIS data derived from different sources. (Map by Brian Tomaszewski.)

Coordinate Systems: The Whole Picture

To summarize the overall ideas of how the various components of a coordinate system work, it is useful to think of a coordinate system in the following way:

1. A horizontal datum (based on a reference ellipsoid and control points) is used to mathematically define the Earth's shape and provide a reference for latitude and longitude (or spherical) coordinates.
2. A map projection mathematically translates a 3D representation of the Earth into a 2D representation, which unavoidably creates some distortion. Based on this translation created through a map projection, spherical coordinates can be converted to planar (x,y) coordinates.
3. A coordinate system can then be derived from an agreed-upon origin point based on map projections that are optimized for a particular region and using standard units of measurement.

The following chapter sections discuss the basic principles of cartography and draw upon the previous discussion as a foundation.

BASIC PRINCIPLES OF CARTOGRAPHY

Maps have existed in human societies as long as there has been recorded history. For example, the Babylonian *Imago Mundi* dates from almost 2,500 years ago and depicts the relationship of the city of Babylon to other cities and surrounding land masses (Figure 2.17).

Figure 2.17 The Imago Mundi, one of the world's oldest surviving maps from ancient Babylonia. Note the cuneiform writing on the top and the graphical depiction of Babylon and several surrounding islands and landmasses. (From British Museum. http://en.wikipedia.org/wiki/File:Babylonianmaps.jpg)

DECIMAL DEGREE COORDINATES

Decimal degree coordinates are a way of referring to latitude and longitude coordinates in numerical/decimal format as opposed to degrees, minutes and seconds. By using a numerical/decimal format, latitude and longitude coordinate pairs, can for example, be entered much easier into a GPS device or used in an online mapping tool like Google Maps. The following example shows how decimal degrees can be calculated from degrees, minutes and seconds.

Background:
1 Degree = 60 Minutes
1 Minute = 60 Seconds
43° 4′ 31″ is the same as 43.0753° How?
Step 1: convert seconds to minutes:
31 seconds/60 = 0.5166 minutes
Step 2: convert minutes to degrees:
4.5166 minutes (or 4 + 0.5166 from step 1.)/60 = 0.0753 degrees
Step 3: Combine the degree value from step 2 onto the original degree or 43°
and 0.0753 = 43.0753

As one the oldest forms of human communication, the art and science of map making, or *cartography*, has evolved to serve countless purposes. Through this evolution, several key principles of cartography have been established. Although the process of making maps has become easier in terms of capabilities that modern GIS can offer, understanding cartographic principles is just as important as ever because it is easy to make poorly designed maps that can mislead and misinform. The following sections discuss the basic principles of cartography in a GIS context so you have a solid foundation to begin making your own maps with GIS.

Mapping Principles

Before discussing specific map types, it is important to understand some basic principles of map construction and design for creating specific maps with GIS for disaster management applications.

Data Measurement

Raw data (Figure 2.18) are measured in four standard ways for map-based presentation. It is important to understand data measurement distinctions as these distinctions have map design choice ramifications, and by extension, how well (or not) the map will be understood. The four standard ways that data are measured include

Nominal: Nominal data involve the assignment of a code to observations in the data, but there is no numerical significance between codes. Nominal data are sometimes referred to as *qualitative* data.

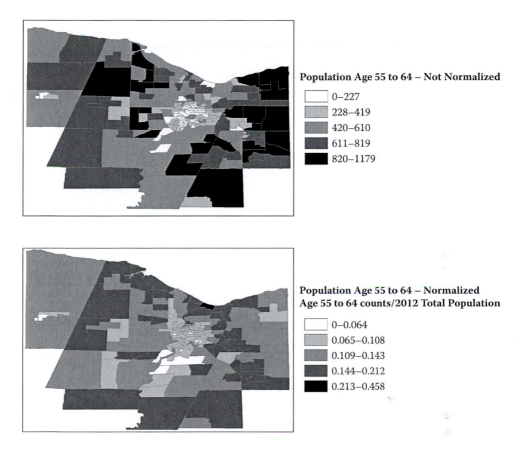

Figure 2.18 Comparison showing raw data counts (top) and the same data presented in normalized format. (Maps by Brian Tomaszewski.)

Ordinal: Ordinal data are data in rank orders (i.e., 1st, 2nd, 3rd) but with no degree of numerical difference between items. A general example of ordinal data would be survey questionnaire responses such as "very good, good, acceptable, poor, and very poor," Although there is a ranking among the responses, there is no indication of what specifically differentiates one category from another (Changing Minds, 2013). In a mapping context, examples of ordinal data were seen in Figure 2.3 in terms of the road networks, which show a ranking of different roads based on the jurisdiction of the roads across federal, state, county, and local authorities. Like nominal data, ordinal data are considered a form of qualitative data.

Interval: Interval data is data that has been ordered with explicit indication of differences between categories based on an arbitrary zero point. The classic example of interval data is temperature. For example, 10 degrees Celsius and 10 degrees Fahrenheit will not feel the same as they use different zero starting points for their measurement. The contour map shown later in Figure 2.27 is an example of interval data mapping. Elevations shown in this map are measured from an arbitrary zero starting point (sea level). Interval data are a form of *quantitative* data measurement.

Ratio: Ratio data are similar to interval data except that there is a nonarbitrary zero starting point as the basis for measurement. Ratio data examples include temperature measured on the Kelvin scale, age, and weight. Ratio is also a form of *quantitative* data measurement.

Visual Variables

Maps are generally created using three basic graphical "building blocks"—points, lines, and areas—in addition to text for labeling map features. From these basic graphical building blocks, data and map feature representation and the message that the data and features are trying to communicate are done through *visual variables*. Visual variables, such as size, shape, orientation, and color hue and lightness are not unique to mapping and are important overall graphical design devices. In a mapping context, they are essential to understand to properly match the correct visual variable with the form of data measurement being mapped. Figure 2.19 shows the ideas of visual variables and their relationship with data measurement using disaster management examples.

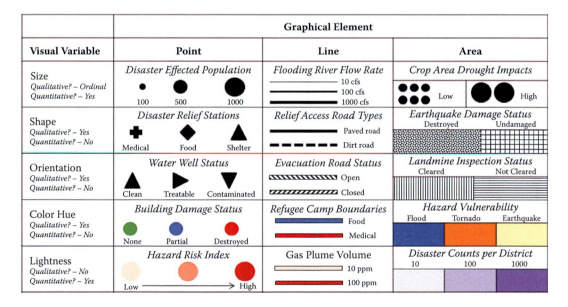

Figure 2.19 A collection of commonly used visual variables with hypothetical disaster mapping examples. Visual variables are powerful graphical devices for communicating messages in map form. However, when designing maps, it is important to remember to match visual variables correctly with data measurement of the feature being mapped. Mismatching visual variables, data measurement, and features can lead to maps that miscommunicate, for example, using different color hues but the same level of lightness to represent quantitative data.

Figure and Ground Relationships

Figure and ground relationships are also important to all forms of graphical design. In a mapping context, *figure* and *ground* refer to the visual display of information such that the elements that are intended to be the map's focus of attention, or the figure(s), are visually contrasted from map elements that provide context, or ground, to the figure elements. Thus, developing effective figure–ground relationships is important for communicating the map's priority message. Figures 2.20a and 2.20b provide disaster management examples of figure–ground relationships.

Figure 2.20a shows a hypothetical disaster example where the point of origin and impact zone of an explosion are displayed in black to make them the figures as they are the most important map features. The surrounding land-use polygons are shown in a light gray and form the ground to provide visual context for the explosion extent figures. Figure 2.20b also shows an example of figure–ground relationships for a hypothetical disaster area map, but in this case, lighter colors are used to establish the figure of the disaster areas and darker colors are used to establish the ground or areas that surround the disaster areas. Both approaches for establishing figure–ground relationships are valid and it is up to the map designer to determine which approach is best, and will ultimately be easily interpreted by the map reader.

Now that you have been exposed to some basic mapping principles, the following sections discuss specific map types that utilize these principles.

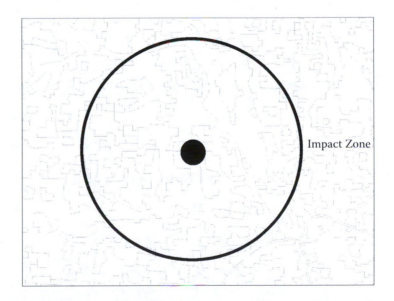

Figure 2.20a Explosion impact map (Maps by Brian Tomaszewski and based on Slocum, Terry A., Robert B. McMaster, Fritz C. Kessler, and Hugh H. Howard. 2008. *Thematic Cartography and Geovisualization*, 3rd edition, Prentice Hall.)

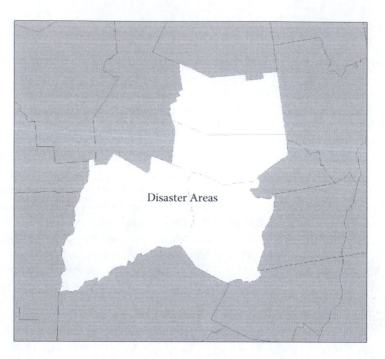

Figure 2.20b Disaster areas map (Map by Brian Tomaszewski and based on Slocum, Terry A., Robert B. McMaster, Fritz C. Kessler, and Hugh H. Howard. 2008. *Thematic Cartography and Geovisualization*, 3rd edition, Prentice Hall.)

Map Types: Reference and Thematic

Maps can be generally classified into two types—reference maps and thematic maps.

Reference Maps

A reference map shows numerous features and does not convey a particular message or communicate specific information. Often in a GIS context, a reference map is referred to as a *base map*, because the base map is the basis from which specific features can be shown. Another way to think of a reference or base map is that it provides the background context to ground features of specific interest that are shown on the map. Common examples of reference maps include the USGS topographic map of the United States or the use of visible satellite imagery in virtual globe technologies such as NASA World Wind or Google Earth™ (Figures 2.21a and 2.21b).

Figure 2.21a shows an excerpt from the West Henrietta USGS 1:24000-scale quadrangle. As a reference map example, note how the map displays several features such as roads, water, contour lines, buildings, and place names. A map like this would be useful for providing geographical context to a spatially oriented activity. As a hypothetical example, if a new water line was going to be added to this area, the USGS map could serve as the base map for showing where the water line would go. Figure 2.21b shows the NASA

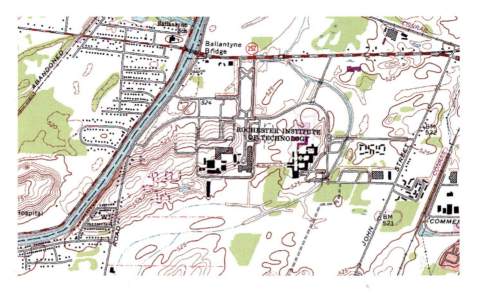

Figure 2.21a   USGS topographic reference map.

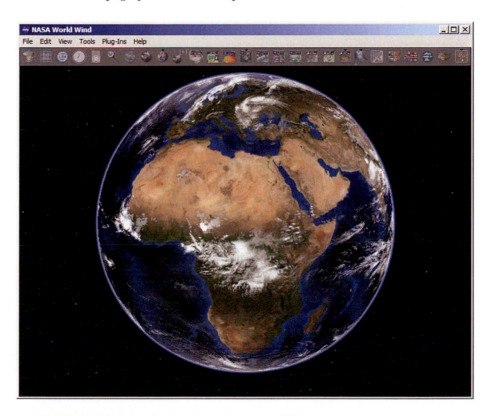

Figure 2.21b   Virtual Globe reference map—Nasa World Wind.

World Wind virtual globe program (http://worldwind.arc.nasa.gov/java/). Virtual globe programs often use visible satellite imagery as a reference map.

An important development in the past five years in terms of the use of reference maps in disaster applications is the advent of free mapping tools that provide "instant" reference mapping capabilities. The most common examples are the Google Maps™ API and OpenStreetMap (Figure 2.22).

As discussed in Chapter 1, web-based technologies such as these are now making mapping capabilities available to a wider audience than previously possible through traditional desktop GIS approaches. Even industry-standard GIS tools such as Environmental

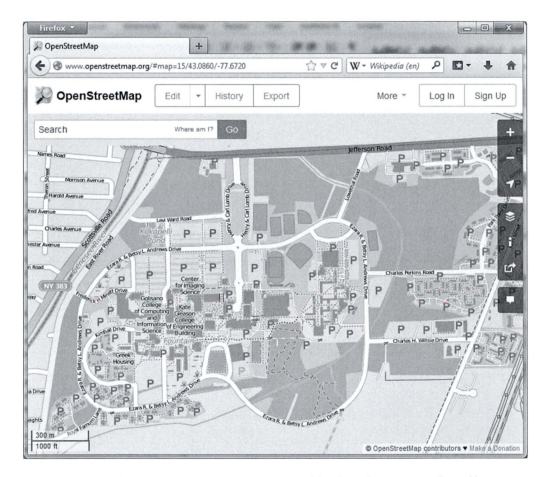

Figure 2.22 OpenStreetMap, a free, open-source, worldwide, reference map (http://www.openstreetmap.org). Note how, like the USGS shown in Figure 2.21a (which shows the same general area as Figure 2.22), OpenStreetMap displays several features such as roads, water, buildings, and place names. With OpenStreetMap, map makers can quickly start mapping features of interest without having to build their own reference map, which is often a very labor-intensive task. OpenStreetMap is discussed further in Chapter 3. (Illustration © OpenStreetMap contributors.)

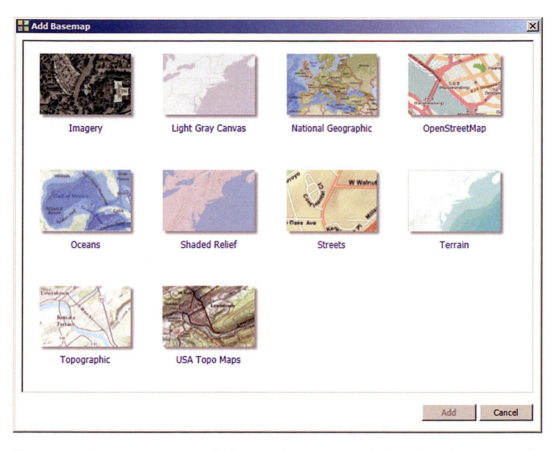

Figure 2.23 Base-map options available in Esri's ArcMap 10.0 desktop GIS software. Note the different varieties of base maps available depending on the mapping needs. (Illustration © 2014 Esri, ArcGIS, ArcMap. All rights reserved. Used with permission.)

Systems Research Institute (Esri) ArcMap now offer a series of reference maps that users can load directly into ArcMap via mapping web services (Figure 2.23).

The advent of instant reference maps is an exciting trend in mapping, but also necessitates consideration of instant reference map ramifications. For example, an instant reference map may not always provide the best design choice as instant reference map displays cannot be modified. As you will see in upcoming sections of this chapter, there are many factors to consider in effective cartographic design and an instant reference map may not be effective for more advanced disaster management mapping needs.

Thematic Maps

Thematic maps convey a specific message—distributions of one or more attributes or relationships among several attributes. They are powerful devices for developing

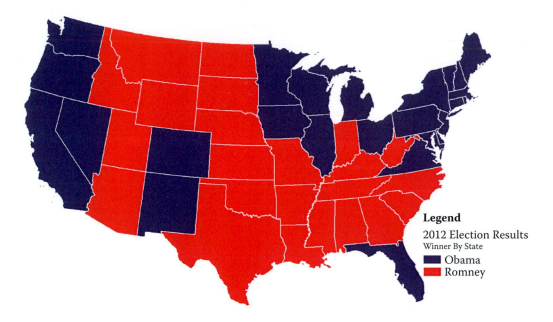

Figure 2.24 A thematic map example; winner-by-state results of the 2012 United States presidential election. In this map, several regional patterns are evident. For example, note the cluster of red states in the southeastern part of the United States that voted for candidate Mitt Romney. Also note the cluster of blue states in the northeastern United States that voted for candidate Barack Obama. Patterns like these can potentially reveal characteristics of the people who live in these regions such as religious or social values. This figure is an example of mapping nominal or qualitative data using different color hues. For example, US states were classified as either having voted for Barack Obama or Mitt Romney. Although these designations were derived from vote counts within each state, there is no *numerical significance* between one state having voted for Obama or Romney. (Map by Brian Tomaszewski with data obtained from *The Guardian*. 2012. Full US 2012 election county-level results to download, Guardian News and Media Limited, http://www. theguardian.com/news/datablog/2012/nov/07/us-2012-election-county-results-download#data [accessed April 2, 2014].)

insights into geographical patterns and trends. You were first introduced to the ideas of thematic maps in Chapter 1, Figure 1.2, which showed the total counts of people aged 65–69 in US counties. Figure 2.24 is another thematic map example that shows a regional pattern.

Thematic maps can also be categorized by the method used to construct the map. The following sections outline specific thematic map categories that are most common for disaster management applications.

Choropleth Maps

A *choropleth map* usually aggregates data for display in a preexisting region such as a state or country. Typically, data are displayed in two ways. The first is with a qualitative distinction between entities such as different color hues to show different land-use types.

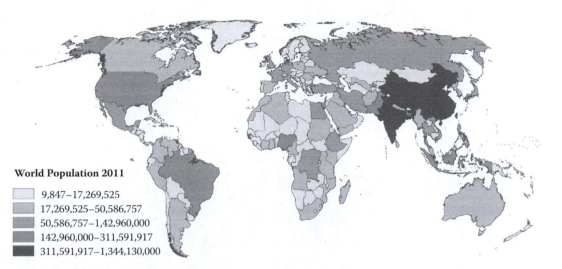

World Population 2011

- 9,847–17,269,525
- 17,269,525–50,586,757
- 50,586,757–1,42,960,000
- 142,960,000–311,591,917
- 311,591,917–1,344,130,000

Figure 2.25 An example of a choropleth thematic map; world population raw counts by country in 2011. Note how lesser populated countries are a lighter shade of blue and as the magnitude of the data (or population) increases, the blue color lightness (or saturation) becomes darker. In a disaster mapping context, a map like this might be a first step in getting a sense of population distributions worldwide and how those populations are vulnerable to disaster impacts. (Map by Brian Tomaszewski with data from World Bank. 2014. Population [Total], World Bank, http://data.worldbank.org/indicator/SP.POP.TOTL/countries?display=default [accessed April 2, 2014].)

The second is with a quantitative distinction where magnitudes of data are shown using different levels of color lightness (or saturation). For example, the darker the color, the greater the magnitude being shown (Figure 2.25).

Since choropleth thematic maps typically show data aggregated based on some region or predefined unit (also known as an *enumeration unit*), it is important to be aware that data aggregation might misinterpret the phenomena being shown. For example, a map showing racial composition of a county based on the highest count of one particular racial group will miss the representation of other racial groups as these others groups will be lost in the aggregation.

Proportional Symbol Maps
Proportional symbol maps use symbols of varying sizes that are proportional to the value or magnitude being shown (Figure 2.26).

Isarithmic Maps
Isarithmic maps use line symbols to display phenomena that are continuous in nature. For example, elevation is continuous; there is never a spot on the Earth's surface that does not have elevation. Thus, contour maps (which are a type of isarithmic map) have been developed to display surface elevation (Figure 2.27).

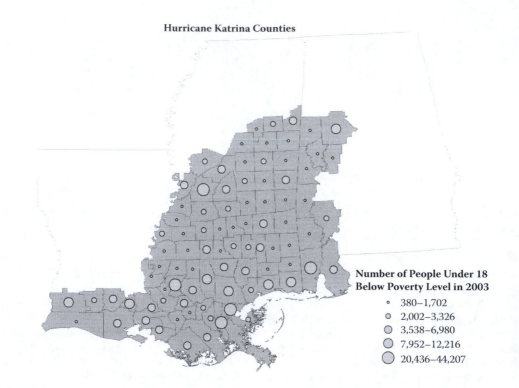

Hurricane Katrina Counties

**Number of People Under 18
Below Poverty Level in 2003**

- ° 380–1,702
- ⊙ 2,002–3,326
- ◯ 3,538–6,980
- ◯ 7,952–12,216
- ◯ 20,436–44,207

Figure 2.26 A proportional symbol map example. This example shows the number of people under the age of 18 living below the poverty level as of 2003 from counties in the Gulf Coast region of the United States eligible for federal disaster assistance after Hurricane Katrina in 2005. A proportional symbol map like this can be useful for comparing differences between counties for disaster vulnerability reduction. (Map by Brian Tomaszewski with data obtained from US Census Bureau. 2012. *Hurricane Katrina*, Census.gov, http://www.census.gov/newsroom/emergencies/hurricane_katrina.html [accessed April 2, 2014], information on Small Area Income and Poverty Estimates program and counties designated by the Federal Emergency Management Agency as eligible to receive individual and public assistance as of September 14, 2005.)

Dot Density Maps

A *dot density map* shows the distribution of an observation or observations at specific points. The basic idea is that each dot can represent one or more instances of the phenomenon at the point, making this a useful technique for showing patterns based on point observations (Figure 2.28).

Summary

Thematic maps are relatively easy to create using GIS tools. However, when creating a thematic or any other map type with a GIS, map design is very important to ensure that the mapped data is not misrepresented or misinterpreted. The following sections present practical map design advice and common mistakes found in GIS-based map making

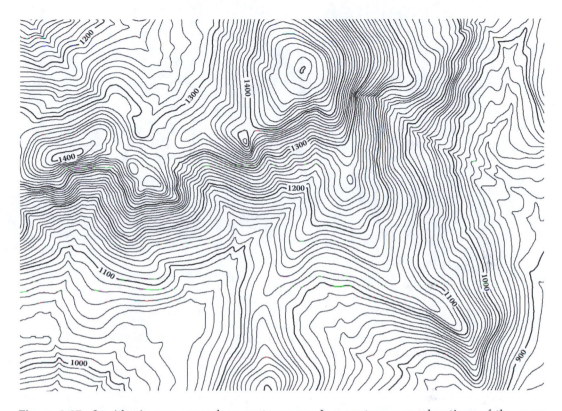

Figure 2.27 Isarithmic map example; a contour map. In a contour map, elevations of the same value are connected using line symbols. In this map, each line represents a 10-meter change in elevation and changes every 100 meters are indicated by elevation labels. The closer the lines are to one another, the greater the elevation increase. A similar approach could be used for other continuous surfaces such as temperature or precipitation. In a disaster management context, a contour map can have multiple uses, such as showing slope gradient, areas susceptible to flooding, or evacuation route planning. (Map by Brian Tomaszewski.)

by new map makers. Developing an understanding and appreciation of these ideas will get you started on making good maps that communicate well and can effectively support disaster management activities.

DESIGNING USABLE MAPS IN A GIS CONTEXT

Making a map that is easy to interpret, understand, and is generally usable is not necessarily difficult to do, but does take practice. Map making in general is often viewed as an iterative process (Figure 2.29).

As shown in Figure 2.29, the process begins with (1) an item from the disaster management cycle (discussed further in Chapter 4) that requires mapping. A disaster mitigation example might be mapping neighborhood flood vulnerabilities. Closely related to the item

Figure 2.28 Dot density map example; worldwide tweets during Hurricane Sandy in 2012. In this example, each black dot represents the location of a Twitter user who revealed his or her location in their Twitter profile and tweeted about Hurricane Sandy between October 28, 2012 and October 31, 2012. As one might expect, the United States appears almost black due to the density of tweets in the United States. However, it is also interesting to note that Western Europe also appears almost black due to the density of tweets, even though the event happened in the United States. Also interesting to note are tweet clusters that appear in other spots around the world such as Africa and South America. With the increased use of social media in disaster management, dot density maps such as this are powerful devices for showing, on a massive scale, instances of individual citizen disaster reporting such as those that can be captured through social media such as Twitter. (Map by Brian Tomaszewski using data obtained through the Twitter API, https://dev.twitter.com/.)

being mapped is (2) the map's audience. Understanding the map audience is important for determining the map's final presentation. Using the previous example of flood mitigation, a map being developed for community members will be different than a flood mitigation map developed for structural engineers or hydrological scientists. After the audience is determined comes the most labor-intensive part in the map making process— (3) collecting data. As you will learn in Chapter 3, GIS can incorporate a very wide range of data that is often modified in some manner such as reprojection or "cleaning" bad data entries. Furthermore, finding data of appropriate scale and detail (as discussed in this chapter) is equally challenging and time consuming. After collecting data, comes development of (4) map representations. Map representations are based on the principles of cartography discussed in this chapter. For example, does the data collected lend itself to some particular thematic map type? What visual variables (and for which map features) should you choose so that your intended audience will understand the map? After developing map representations and developing a final map product, the map is then presented to (5) the map audience from which feedback is obtained, and in turn, changes made to the map in a iterative cycle (as shown by the line/arrows in Figure 2.29). Using the running

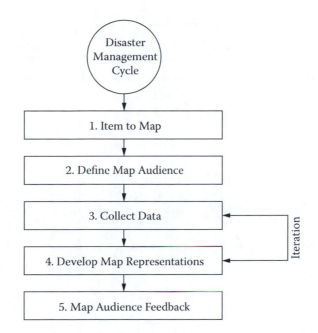

Figure 2.29 The map-making process. (Adapted from Slocum, Terry A., Robert B. McMaster, Fritz C. Kessler, and Hugh H. Howard. 2008. *Thematic Cartography and Geovisualization*, 3rd edition, Prentice Hall.)

example of a flood mitigation map to illustrate these points, community members might request that a different legend be used so the map is easier to understand or hydrological scientists may request that additional items be added to the map to support different types of scientific inquiry.

It is also useful to think of the map-making process as a dialogue. For example, "We need to create a map of people who are vulnerable to coastal floods in the eastern United States for FEMA decision makers. Data will be collected from US census figures on elderly and disabled people and will be projected into an equal area projection to not distort areas shown. A choropleth map using different color lightness will be used to display normalized numbers of elderly and disabled people as a percentage of overall census tract population. We anticipate that once the map is developed, additional vulnerability dimensions, such as lower-income families, will be included."

Common Examples of Poorly Made Maps Created with a GIS

It is beyond the scope of this chapter to provide detailed discussion of effective cartographic design. See the Resources section of this chapter for this type of information. What follows next, though, are common examples of poorly made maps created using an industry standard desktop GIS tool (Figures 2.30–2.35). These examples are drawn from real novice GIS students and demonstrate how GIS makes it easy to create a bad map. Study and refer back to these figures if you are new to GIS map making.

DEC_Road_and_Trails

dec_land

Figure 2.30   Common legend issue 1: remove_underscores_from_legend_items_as_they_look_funny.

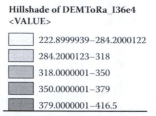

Hillshade of DEMToRa_I36e4
<VALUE>

222.8999939–284.2000122

284.2000123–318

318.0000001–350

350.0000001–379

379.0000001–416.5

Figure 2.31   Common legend issue 2: Round numbers in legend items so there aren't as many decimal places.

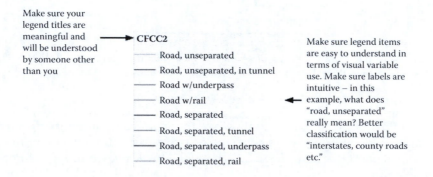

Make sure your legend titles are meaningful and will be understood by someone other than you

CFCC2

Road, unseparated
Road, unseparated, in tunnel
Road w/underpass
Road w/rail
Road, separated
Road, separated, tunnel
Road, separated, underpass
Road, separated, rail

Make sure legend items are easy to understand in terms of visual variable use. Make sure labels are intuitive – in this example, what does "road, unseparated" really mean? Better classification would be "interstates, county roads etc."

Figure 2.32   Common legend issue 3: Make legend items clear and easy to understand.

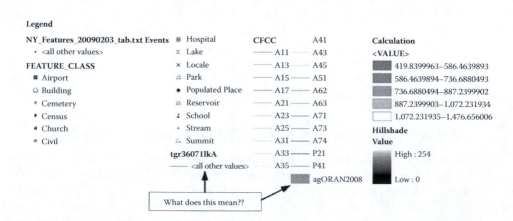

Legend

NY_Features_20090203_tab.txt Events
• <all other values>
FEATURE_CLASS
▪ Airport
▢ Building
▴ Cemetery
✦ Census
◂ Church
✷ Civil

▨ Hospital
⊏ Lake
× Locale
⇄ Park
● Populated Place
♫ Reservoir
♪ School
⁎ Stream
⏃ Summit
tgr36071lkA
<all other values>

CFCC
A11
A13
A15
A17
A21
A23
A25
A31
A33
A35

A41
A43
A45
A51
A62
A63
A71
A73
A74
P21
P41
agORAN2008

Calculation
<VALUE>
419.8399963–586.4639893
586.4639894–736.6880493
736.6880494–887.2399902
887.2399903–1,072.231934
1,072.231935–1,476.656006
**Hillshade
Value**
High : 254
Low : 0

What does this mean??

Figure 2.33   Common legend issue 4: Make sure all legend items are relevant.

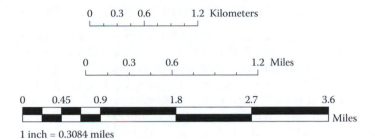

Figure 2.34 Common scale bar issue: Use even rounded numbers for scale bar increments. For example, end the graphic scale at 5 miles or 2.5 miles.

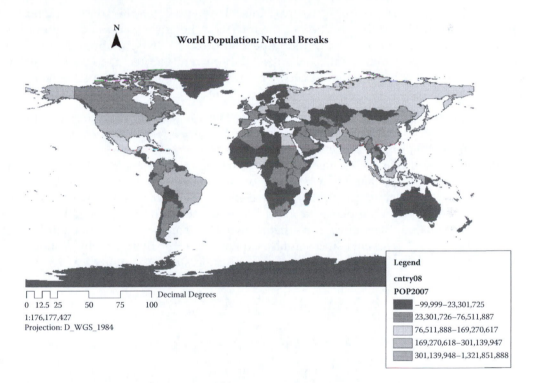

Figure 2.35 Suboptimal quantitative data. In this map (which shows world population by countries in 2007), make note of how different color hues are being used to show population magnitudes. This map uses *diverging color schemes*, which can be used for equal emphasis on mid-range and extreme data range end values (From Brewer, Cynthia A. 1994. Color use guidelines for mapping and visualization. *Visualization in Modern Cartography* 2:123–148). However, the map above fails to effectively represent mid-range values. It difficult (if not impossible) to clearly discern mid-range values due to lack of change in lightness values as the yellow is as visually prominent as the other colors in terms of lightness. A better way to present this type of data would be by varying the color lightness of a single hue (i.e., from dark blue to light blue). Also make note of some of the legend issues outlined previously.

INTERVIEW WITH DR. ANTHONY C. ROBINSON

Dr. Anthony C. Robinson (Figure 2.36) is the faculty lead for Online Geospatial Education for the John A. Dutton e-Education Institute and assistant director of the GeoVISTA research center in the Department of Geography at Pennsylvania State University.

As an internationally recognized cartographic scholar and geographic information scientist, Dr. Robinson's research focuses on the science of interface and interaction design for geovisualization and geovisual analytics tools, the design of map symbol standards, developing tools for collecting and adding meaning to geographic information, and eye-tracking to design new geovisualization techniques. In 2013, he was the instructor of *Maps and the Geospatial Revolution*, a massive open online course (MOOC) that drew more than 40,000 students from around the world interested in the topic of maps and mapping. He holds a PhD in geography from Penn State.

The following is the first of a two-part interview with Dr. Robinson conducted for this book in June 2013. In this portion of the interview, he answers questions about cartographic needs during disasters, thinking beyond crisis response mapping, and cartographic design opportunities and challenges with current GIS technology. The second part of this interview is presented in Chapter 9, where Dr. Robinson discusses the future of disaster mapping.

With more and more people turning to tools like Google Maps for mapping needs during a disaster, how important do you think knowledge of cartography is for people interested in making maps to support disaster management activities?

I think it is really important for people who are trying to make their own maps for the first time. Now that we have these great affordances for doing that, we also need them to understand that there is a science that underpins how we design representations of the planet to make sense to people. In order for folks to move beyond simply making maps that show where something is—for example, the location of an event—and actually explain why it is there or why it should be there or why it should not be there, to understand the analytical reason for

Figure 2.36 Dr. Anthony Robinson.

having a certain thing happen, I think you have to understand how to show that kind of stuff to laypeople. So the reason we have to teach map design is because it is not common sense all the time and I see a lot of map artifacts now that are visually very cluttered, they use colors schemes that are not appropriate to the data types that they are trying to show, or they make assumptions about causation which may not actually mean anything if they haven't normalized the data or accounted for those variables appropriately. So I think there is an opportunity in that we have a lot of people making maps now for the first time, which is great, but I think there is an opportunity for those of us who work on the academic side of teaching about maps to develop better frameworks, tools and examples to help people tell stories with maps and rather than just showing the presence and absence of things.

Are you seeing in your own teaching experience people that were focused on Google Maps the more online tools make progress with learning core cartographic design?

I think it's creeping in a little bit. The example I think of right off the bat is how Ushahidi has evolved a fair bit since it began. It's not using Google Maps as an underpinning, but it's still using an OSM [OpenStreetMap]-based base map. The design aesthetic is based on the kind of map you would use for navigating, but now there are symbols and such on top that are aggregating automatically [see Figure 1.8a and b from Chapter 1 for an example of this] when you reach certain scales so you can avoid clutter issues. There are still issues they have with colors and normalization and they don't have really any analytical tools to help you predict or understand if there are clusters that are significant, but the map design itself does seem to have evolved from the simple pins on the map that it originally had.

I think it has reached a critical mass now where the crisis management community still believes that simple mashups are what you need—that a map of a crisis is a bunch of dots showing where the buildings [are] that are damaged. I think that those of us that understand geography and geographic analysis are right to say, "well, actually you need more than that, it's not just the presence and absence of things that is important, that's just one part of it," and really we would see that as an input to an analytical product that *explains* a pattern that's different or noteworthy. I think that's the goal we want to try to achieve next is to make it possible to understand and explain a situation using a map rather than just saying "here is where things are"; we need to get beyond that.

Disaster response often gets the most attention from the media. How do you think mapping, and in particular cartographic design, can support other aspects of disaster management such as disaster recovery and mitigation?

That's a really good point. I think you're right that people pay attention to response maps like Ushahidi, and I hope that the folks that develop those kinds of systems will also start to develop longer-term systems that support recovery processes, though that will be really hard! I'm not sure I have an answer as to how you make an interactive web-map compel people to continue in engaging in acts of mitigation over a long period of time—that's a really difficult thing to try to do. I'm pretty sure there has been some effort from the gulf oil spill to do that because the very nature of that disaster sort of suggests this much longer-term need that making sure that the coast line has adequate surveillance in case more oil pops up. They're still finding oil in Alaska from the *Exxon Valdez*, so I think that particular type of disaster seems to have gotten some traction from people looking at long-term monitoring, but it's a way bigger challenge to keep people interested, especially these things that are really based on volunteered geographic information [VGI]. How do you motivate volunteers to keep continue contributing to a map a year after a disaster happened? For example, are there any examples right now that we can find of people doing long-term recovery using geography as an interface for Haiti? I would be surprised if we could find one. I would look to the agencies that fund that longer-term redevelopment stuff like UN-OCHA [Office for the Coordination of Humanitarian Affairs] to establish a set of tasks that they think are really important for those longer-term recovery efforts. Then I think it is up to the academic cartography community and the business community that develops these technologies to take those tasks and actually try to run with them. And I think a good approach would be for an organization like ISCRAM [Information Systems for Crisis Response and Management] to create a scenario and challenge teams of academics or business partners to come up with solutions that are not for response (which is what normally gets built for those organizations), but to actually build innovative stuff that may work to engage people on long-term issues of response and recovery. I think that would be a cool way to try to encourage innovation in that area. It's a real challenge because even if you were successful and you had what you wanted to know about how public risk perception changed since a hurricane hit a particular coastal community. Maybe you want to know whether or not those people are likely to evacuate next time when they should have done it last time. What if you were able to get 5,000 qualitative responses to that question? You could argue that you could make a pretty cool story then about all these things that are happening, but I would challenge you to show that on a map—it would be really hard to do.

Would this get more into advanced things like spatialization of data, perhaps a geographic map isn't the best thing but some kind of spatial representation of words or trends?

Yes, and I think that is where there are some open research questions still about what kind of approach would work. Should we just be linking a map view to that qualitative information in two different windows? Or are coupled, hybridized views that show the map in the main interface, and you can still access that kind of qualitative information somewhat efficiently inside the map itself? I think there is a real challenge in showing an overview. How do you make the overview of the kinds of things that are going to be the rich types of feedback that you have during a recovery and mitigation phase? Let's say you have plans for proposed building codes that you want people to debate about. The feedback you want from that is basically all qualitative—you're not going to be voting yes or no. They're going to say "yes, but … I think the minimum building height should be this and here's why … and yes, but you can't make a school on stilts." We want to be able to capture those perspectives because that's how recovery and mitigation happens. I think there is a real research question to be answered there about how can we show that in one fully integrated display or do we have to use spatializations and hybrid views that are linked to each other—which one is more effective for actually telling these stories? If we make progress on figuring out how to show social media, then that kind of stuff will naturally fit well with these other kinds of narratives, and there are a lot of people of working on that. If you can make sense, visually, out of 10,000 tweets, and I'm talking about the content inside those tweets, if you're able to do that on a map someday, then that same capability ought to work pretty well for a somewhat richer narrative—but we haven't solved that first problem yet.

What issues/problems/challenges (if any) do you see with incorporating good cartographic design principles into disaster mapping?

A lot of things have to do with understanding what kind of data are you trying to show. Cartographers usually start with three major problems—one is, what audience are you trying to talk to? The other problem is what format do you to have to talk to those people and then what is the purpose of your message? What are you trying to communicate? So we think through those problems before designing a map. It's difficult to imagine digital mapping systems having the ability to translate a nonexpert view of a situation to provide answers to those questions, and our current systems certainly don't do a good job with that. They don't help people by saying, "Hey, what are you trying to show? Who are you showing it to? What's the purpose, etc.?" Instead, we have systems that essentially say, "OK, you can do anything you want! Go ahead, make a map!" It's kind of like if people were using word processing software before knowing how to write a sentence. We need to engineer the "map machine"—the software that makes the maps—to ask these questions in a clever way to the end users so that they can make better design decisions. One example that I think is a good one to look at in terms of how this might work, in terms of recommending to users how to make design decisions, would be to note how Geocommons [http://geocommons.com/] works. When you select a dataset, it starts to suggest to you based on what you have and the classification methods that are appropriate for that kind of data. And it teaches you right in the interface how to recognize that stuff yourself. It is a good example to look at because it does suggest to the end user, who may not know anything about cartography, "Hey, here are good color schemes to use because you have categorical data." I hope to see more GIS using that kind of smart-assistant approach to help people make wiser decisions about symbology and colors and other map design directions. It's not going to solve everything, but it may teach people along the way about simple stuff like using qualitative colors only when you have categorical data—some really obvious pitfalls. Another one of the major challenges that we have with incorporating design principles into disaster mapping is that the people that are best equipped to make nicely designed maps tend to take a lot of time to do that. You put a lot of effort into making one really nice information graphic and you don't have time in a disaster to do that. And you may not have people in the loop there that have any cartographic training, which is where those smart systems come into play or perhaps really well-designed templates could be beneficial—although they're not likely to always match what people need in a crisis situation where the nature of the beast is that you can't predict what's going to be useful. So I think there is an issue to overcome too with how we get people up to speed quickly to learn how to make effective graphics. I haven't seen anything yet that would be like a graphical commons, where you could leverage volunteers who actually know how to do this stuff to become part of the workflow in a crisis situation and be looking at the graphics and other media generated for place and for immediate and obvious improvement. What if you could leverage volunteered resources in terms of volunteered expertise in cartography or graphic design as well as we've been doing with other things? I think that would be a cool direction to move, and I hope that folks in the crisis mappers community, for example, take that up as one of their initiatives.

How well do you think current desktop GIS (that GIS professionals use) are at supporting good cartographic design and is this support sufficient for disaster management mapping needs?

I think current desktop systems have capabilities that make it possible to make appropriate, good-looking maps that are interpretable by lots of people. However, the affordances for making those design decisions are a lot worse than they are in dedicated graphic design software packages. So while it is possible to change, let's say, the typeface used for every label on a map in a desktop GIS, even the way that's rendered on screen isn't as nice as in a graphic design tool. The possible area for improvement is that on the flipside of that is that some GIS systems are very good at applying templates. You can give them any kind of data and it will try to symbolize it in a certain way. So, it's not impossible to imagine some of the cartography focus to be on designing really effective templates that embody best practices, and that those could be used to help streamline the process of map design. Once again, the time issue is really important here—how much time is there in a crisis situation for someone to focus on changing all of the labels to make sure that every one of them is visible and not in conflict with anything else on the screen? That's probably something that no one will have time to do in an emergency situation. There are other simple things, such as if you need to create a series of choropleth maps using the same variables across different years, how do you quickly apply the same classification scheme to all of them? At the moment, you have to manually go in and apply those class breaks and it's a tedious, error-prone process. A lot people just won't do it. So you run the risk of having maps that look like they are in a series maybe, perhaps use the same color schemes, but which have different class boundaries, making them not comparable. That's how you can end up with people making poor decisions about geographic data. So I think there's a lot of room for improvement when it comes to making sure that good cartographic design is just as easy to execute, and as fast and efficient to execute as it is to transform datasets and do spatial analysis operations. We can move pretty quickly now that we have cloud resources; you can even compute on a very large scale if you need to and have that capability on demand. What we can't do yet is to design in the cloud. That's the analogy I would try to use, but it's a little imperfect. Imagine if you could scale the process of designing maps and information graphics. We're not able to say, "OK, this map is not really good because the aggregated units it is using are not helping us tell the story." Those are decisions that are really hard to automate, but I think that is the goal we need to have—just like we can scale up computing now, we have to be able scale up design to meet sudden and acute demand for high-quality, designed maps and graphics to communicate complex concepts.

CHAPTER SUMMARY

You learned about the fundamentals of geographic information and maps in this chapter. You were first introduced to how data is different than information and the importance for understanding these distinctions in a GIS context. This discussion was followed by one of the most basic and important concepts in maps and mapping—scale. You learned that there are three ways of representing scale, the difference between large- and small-scale maps, and why scale is important in terms of detail and accuracy.

Map projections were then discussed as methods for representing the earth's three-dimensional shape in a two-dimensional representation. You were shown how map projections make trade-offs in terms of shape, size, area, distance, and angle, and that many map projection classes have been developed by cartographers to account for these trade-offs.

Coordinate systems, which are based on map projection classes, were then discussed. You were shown how planar coordinate systems help to reference locations on the earth's surface using Cartesian x,y coordinates as opposed to spherical latitude and longitude coordinates, which use angular measurements to reference locations. Both coordinate system types, however, use a datum that accounts for the Earth's shape and uses well-known control points for referencing and indexing locations. The principles of cartography then followed.

You learned about different ways in which data is measured in terms of qualitative and quantitative data measurement. You were shown specific visual variable examples. Visual variables are important visual devices for representing map features. You also learned about and were shown how specific examples of figure–ground relationships are important for the visual structuring of map features. You were then shown specific map type examples. You first learned about reference maps, which have no specific message, and were shown examples of reference maps found in popular online mapping tools. Next, you were shown examples a four specific types of thematic maps, or maps designed to have a specific purpose or message. Disaster thematic mapping examples were provided to show you how the previously discussed ideas of data measurement, visual variables, and figure–ground relationships converge when creating thematic maps. The chapter ended with some practical advice on how you can create usable maps with GIS. You were shown a framework for the map design process that you can follow once you begin to make maps. You were then shown some specific, common problems that new map makers make when first learning to create maps with GIS. Finally, an interview with one of the world's leading cartographic thinkers and educators provided you with some important ideas to consider about disaster mapping. In the next chapter, Geographic Information Systems are formally discussed in terms of specific GIS data formats, basic functions of GIS agnostic of any particular GIS software product (analysis, map production/cartography, data modeling), followed by an overview of specific commercial and free and open-source GIS software products relevant to disaster management.

DISCUSSION QUESTIONS

1. What scale would you approximately need to use for the following types of emergencies and disasters: (a) a neighborhood blackout, (b) a snow storm affecting a small (<250,000 people) city, (c) a hurricane hitting the east coast of the United States, and (d) a major tsunami in the western Pacific Ocean?
2. What types of classes of map projections might you use for the scenarios listed in question 1 and why?
3. What types of disaster situations would lend themselves to choropleth mapping?
4. How might you combine different visual variables for multivariable mapping. For example, size and lightness?
5. Suppose you have to make a disaster map and differing color hue is not an option due to lack of having a color printer available. How would you reconsider the use of visual variables?
6. Referring back to Figure 2.33 (common legend issue 4), what other issues can you find with this legend?

RESOURCES

The following is a nonexhaustive list of reading that can provide you with further information on topics discussed in this chapter.

Principles of Mapping

D. DiBiase, J. Sloan, W. Stroh, and B. King. *Nature of Geographic Information*. Penn State, College of Earth and Mineral Sciences, Department of Geology, 2011, https://www.e-education.psu.edu/natureofgeoinfo/.

Geodesy (including Datums and Reference Ellipsoids)

J. Müller and W. Torge. *Geodesy*, 4th edition. DeGruyter, Berlin, Germany, 2012.

History of Cartography

The History of Cartography series. 1987–. University of Chicago Press, 1987–, http://www.press.uchicago.edu/books/HOC/index.html.

Basics of Statistical Data Classification for Maps

T. A. Slocum, R. B. McMaster, F. C. Kessler, and H. H. Howard, *Thematic Cartography and Geovisualization*, 3rd edition. Prentice Hall, Upper Saddle River, New Jersey, 2008.

Designing Good Maps in a GIS Context

C. Brewer, *Designing Better Maps: A Guide for GIS Users*. Environmental Systems Research, 2004.
G. Dailey, *Normalizing Census Data Using ArcMap*, http://www.esri.com/news/arcuser/0206/files/normalize2.pdf.

Map Color

ColorBrewer 2.0: Color Advice for Cartography, http://colorbrewer2.org/. The ColorBrewer tool was used to select colors used in all of this book's color maps.

REFERENCES

Brewer, Cynthia A. 1994. "Color use guidelines for mapping and visualization." *Visualization in Modern Cartography* 2:123–148.

Changing Minds. 2013. "Types of data," Changing Minds.org, http://changingminds.org/explanations/research/measurement/types_data.htm (accessed April 2, 2014).

Environmental Systems Research Institute. 2010. "What is a map projection?" ArcGIS Desktop 9.3 Help, http://webhelp.esri.com/arcgisdesktop/9.3/index.cfm?topicname=what_is_a_map_projection? (accessed April 2, 2014).

Furuti, Carlos A. 2008. "Azimuthal projections," http://www.progonos.com/furuti/MapProj/Normal/ProjAz/projAz.html (accessed April 2, 2014).

The Guardian. 2012. "Full US 2012 election county-level results to download," Guardian News and Media Limited, http://www.theguardian.com/news/datablog/2012/nov/07/us-2012-election-county-results-download#data (accessed April 2, 2014).

National Atlas of the United States. 2013. "Map projections: From spherical earth to flat map," National Atlas.gov, http://www.nationalatlas.gov/articles/mapping/a_projections.html (accessed April 2, 2014).

National Oceanographic Partnership Program (NOPP). n.d. "Track a NOPP Drifter," http://drifters.doe.gov/track-a-yoto/track-a-drifter.html (accessed April 2, 2014).

Robinson, Arthur H., Joel L. Morrison, Phillip C. Muehrcke, A. Jon Kimerling, and Stephen C. Guptill. 1995. *Elements of Cartography*, 6th edition, John Wiley & Sons. Inc., Hoboken, New Jersey.

Slocum, Terry A., Robert B. McMaster, Fritz C. Kessler, and Hugh H. Howard. 2008. *Thematic Cartography and Geovisualization*, 3rd edition, Prentice Hall, Upper Saddle River, New Jersey.

US Census Bureau. 2012. "Hurricane Katrina," Census.gov, http://www.census.gov/newsroom/emergencies/hurricane_katrina.html (accessed April 2, 2014).

US Geological Survey. 2006. "Map Accuracy Standards: Fact Sheet FS-171-99," USGS, http://egsc.usgs.gov/isb/pubs/factsheets/fs17199.html (accessed April 17, 2014).

World Bank. 2014. "Population (Total)," World Bank, http://data.worldbank.org/indicator/SP.POP.TOTL/countries?display=default (accessed April 2, 2014).

3

Geographic Information Systems

CHAPTER OBJECTIVES

Upon chapter completion, readers should be able to

1. understand the components of GIS,
2. understand the concept of layers in GIS,
3. be familiar with common GIS functions and how they relate to disaster management,
4. be familiar with common GIS data storage formats,
5. understand some of the limitations of GIS,
6. understand what GIS metadata is and why it is important, and
7. identify and discern a variety of specific GIS technologies that are relevant to disaster management.

INTRODUCTION

This chapter formally presents Geographic Information Systems (GIS). The chapter starts with a discussion of what a GIS is and what it can and cannot do. In this discussion, specific GIS software product references are avoided as much as a possible to ensure you understand the underlying principles of GIS that are used in specific GIS technologies. The chapter then discusses the most important and time-consuming aspect of working with a GIS—GIS data. You will learn about conceptual differences in how the earth is digitally referenced and represented in various GIS data model formats. Additionally, you will learn about GIS *metadata*—a critical item to ensure that the GIS data you select is relevant to your GIS operational needs. The second half of the chapter is devoted to specific GIS technology. In this part of the chapter, you will learn about commercial, open-source, and open/web-based GIS software. This part of the book is the one that is most likely to change quickly as specific GIS technology is constantly and rapidly changing, and you will need to keep track of these changes. However, the foundational concepts learned in the beginning of this chapter should serve to keep your skill and knowledge set relevant even as specific GIS software products change. The first section presents an overview of GIS.

WHAT IS GIS?

To understand what a GIS is, let's first take a closer look at what the acronym means. Remove the "G" from GIS and you have *IS* or an *information system.* Information systems have been defined as "combinations of hardware, software, and telecommunications networks that people build and use to collect, create, and distribute useful data, typically in organizational settings" (Valacich and Schneider, 2010, 8). Additionally, like any system that is a whole constructed of parts, a GIS can also be viewed as an amalgam of several parts that create the overall system. Although variations exist in what exactly those parts might be, a GIS is generally considered to be composed of the following interrelated parts that follow the information systems definition closely:

1. **Software:** Software is used for running GIS operations. For example, commercial GIS software packages such as Esri's ArcMap or open-source web mapping environments such as Open Layers.
2. **Hardware:** Hardware is the platform in which software is run and/or data is stored. In today's increasingly interconnected world, hardware can range from traditional PCs to smartphones to massive computing infrastructures for hosting cloud computing resources.
3. **People:** People include those who work with GIS in a variety of capacities such as using GIS to make decisions, organizations like the United Nations Geographical Information Working Group (UNGIWG - http://www.ungiwg.org/) that advocate for GIS, or students learning about GIS.
4. **Knowledge:** Knowledge is perhaps the most abstract part of GIS but as equally important as the other parts. Knowledge, in the context of this discussion, refers to the variety of training, education, skills, and experience that are applicable to GIS. For example, by reading this book, you are gaining new knowledge in GIS, cartography, spatial analysis, and spatial perspectives on how these and other ideas in this book are applied to the disaster management domain.
5. **Data:** Data will always be the more important component of a GIS. Representation of the earth's features, which is the conceptual core of GIS, is fundamentally based on data and hence why significant discussion of GIS data is made in this chapter.
6. **Network:** The network can be considered the element that connects all the other parts together. For example, the Internet that connects people to GIS data websites or connecting GIS software with web-based data services or social networks that connect people who use GIS with one another through things like GIS user communities.

Figure 3.1 depicts a graphical representation of the components of GIS with disaster management items added to illustrate the components in the context of GIS for disaster management.

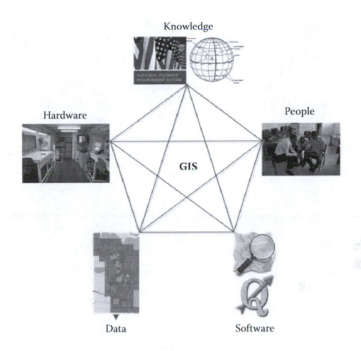

Figure 3.1 The components of GIS. The lines represent networks that connect all of the other components. Each component requires elements of the other components. For example, the *hardware* in the GIS truck requires *people* with *knowledge* of disaster management practice and fundamentals of geographic information to use *software* that consumes *data* to support disaster management activities. (Illustration © OpenStreetMap contributors.)

A Brief History of GIS

Although maps have existed for millennia, the origins of what we now consider GIS are usually attributed to what actually might be considered a disaster (or emergency) management scenario. During the London cholera outbreak of 1854, physician Dr. John Snow famously mapped cholera outbreak instances to find spatial clusters that led him to the conclusion that cholera was originating from a contaminated well (Figure 3.2).

The 1960s saw the beginning of the development of modern GIS. Creation of the term *Geographic Information Systems* is credited to Dr. Roger Tomlinson (1933–2014) and development of the Canadian Geographic Information System, which was a first step in moving beyond computer mapping to include map layer overlays. The later 1960s and early 1970s were also when the Harvard Laboratory for Computer Graphics and Analysis began developing some of the first computer-based spatial analysis and computer cartography and graphics research and applications, and later, the ODYSSEY system, which was designed to process larger geographic datasets (Chrisman, n.d.). Jack Dangermond, president and founder of Esri, began his career at the Harvard Laboratory for Computer Graphics and Analysis. By the early 1980s, Esri was founded and the ArcINFO software product running on UNIX platforms was developed for specialized application uses such as environmental management and demography via the Dual Independent Map Encoding (DIME), and later led to the Topologically Integrated Geographic Encoding and Referencing (TIGER) formats developed by the US Census Bureau, of which TIGER is still used today. By the 1990s, GIS technology then began a trend that continues to this day of closely following and being shaped by broader computing industry trends. For example, the 1990s saw the first graphical user interface (GUI) and mainstreaming of GIS technology to align with the rise of the PC and Windows, the 2000s saw increased use of web-based and Internet GIS and the time when the first edition of this book was written (2014) is currently the age of cloud computing, mobile computing, social media, and "big" datasets.

Figure 3.2 Excerpt of John Snow's famous 1854 map of cholera outbreaks. The cluster of cholera cases found near the pump of Broad Street (seen in the center of this image) led to the conclusion that this pump was the cholera source. (From John Snow, *On the Mode of Communication of Cholera*, 2nd ed., John Churchill, New Burlington Street, London, England, 1855.)

Organizing the World Geographically: Map Layers

The core power of GIS is its ability to organize data into one common geographic view. This simple statement may seem self-evident given the extensive discussion so far about maps and the principals of geographic information such as coordinate systems that provide a common geographic index. However, the key thing that GIS provides to the organization of data geographically is the concept of map layers. Figure 3.3 provides a graphical representation of the concept of map layers from the perspective of disaster management.

In Figure 3.3, a selection of real GIS datasets from Manhattan, New York during Hurricane Sandy (2012) are shown to demonstrate how map layers are combined to support disaster management. Furthermore, the category of map data (reference vs. thematic) as per Chapter 2 discussions, are also shown to give you a sense of how different kinds of map layers are combined. For example, imagery provides a visual reference to the geographic region in question; the census tracks layer shows population thematic characteristics, tax parcels, and who owns what buildings; the roads layer provides reference to critical infrastructure; the social media layer represents locations of people who are tweeting about the hurricane; and the hospitals layer provides reference for medical issues.

The concept of map layers is itself not a new idea, as acetate map overlays existed for years before the advent of computers. What makes modern GIS-driven map layers so powerful is the ability to overlay any number of digital map layers together and reference them to a common geography, thus allowing for entities on the layers to be viewed and analyzed together with the interactive power that GIS offers, such as quickly changing the map layer, symbology of the map layer, or any of the other GIS functions discussed later in this chapter.

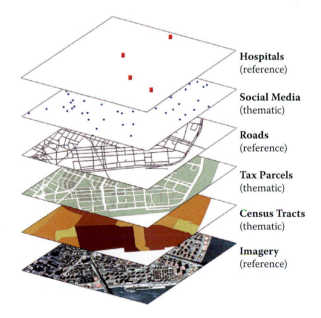

Hospitals
(reference)

Social Media
(thematic)

Roads
(reference)

Tax Parcels
(thematic)

Census Tracts
(thematic)

Imagery
(reference)

Figure 3.3 The layering of geographic information.

What Can You Do (and Not Do) with GIS Software?

GIS software contains many powerful tools that can serve numerous functions. The following sections discuss some of the important GIS functions within the context of disaster management.

Data and Spatial Asset Management

As discussed previously, data is the most important component within the overall system that is GIS. The management of data using GIS is thus a primary GIS function. Management of GIS data can come in many forms. For example, GIS is often used to *create* spatially referenced data. Creation of spatial data can involve many activities such as digitizing features from images (Figure 3.4).

In Figure 3.4, features from an area flooded after the 2011 Fukashima Tsunami in Japan are being digitized from a satellite image of the disaster zone. As can be seen in the middle-left of the image, a variety of construction tools such as Polygon, Rectangle, Circle, and others are available to create features in two categories—flooded areas and standing structures (seen in Figure 3.4 in the construction tools). On the image itself, flooded area polygons have been digitized (or traced) from the image and are shown as polygons with a slanted line fill and standing structures are shown as black-filled polygons. Digitizing features from satellite images that show a disaster impact is a very common technique used to create damage assessment reports such as amount of area flooded and buildings that are intact. Once GIS data are created (or while it is being created), it must be stored in some type of data repository so that it can later be queried, retrieved, disseminated, and updated. Data repositories for GIS data are as diverse as the GIS data itself.

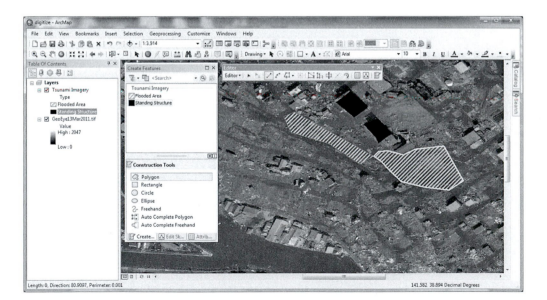

Figure 3.4 Digitizing features from images from the Fukashima, Japan, disaster of 2011. (Copyright © 2014 Esri, ArcGIS, ArcMap. All rights reserved. Used with permission.)

Figure 3.5 show one of the most basic yet still commonly used GIS data storage formats—the comma-separated values or CSV file.

CSV files are nothing more than ASCII-based text files where data in the file is structured using commas (,) to define data columns and each line in the file represents a single data record. Most often used for storing point features, specific geographic information is often represented as decimal degree, *x,y* coordinate numbers in the file that can then be parsed or read by GIS software for rendering on a map. CSV files are a common data storage format used by GIS data providers such as the US Census Bureau and the United States Geological Survey (USGS). Additionally, other characters such as a pipe (|) or tab can be used to structure text-based data like a .csv file.

Figure 3.6 shows a shapefile.

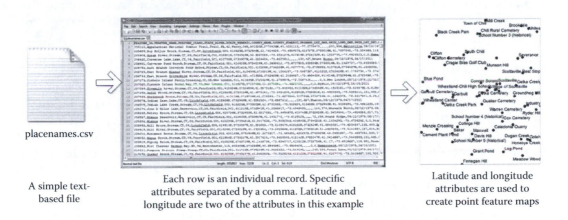

placenames.csv

A simple text-based file

Each row is an individual record. Specific attributes separated by a comma. Latitude and longitude are two of the attributes in this example

Latitude and longitude attributes are used to create point feature maps

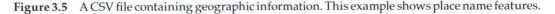

Figure 3.5　A CSV file containing geographic information. This example shows place name features.

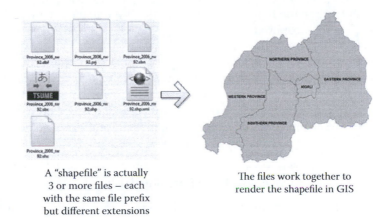

A "shapefile" is actually 3 or more files – each with the same file prefix but different extensions

The files work together to render the shapefile in GIS

Figure 3.6　An example of the specific files that comprise a shapefile and what the contents of a shapefile look like when displayed in GIS, using Rwandan provinces as an example.

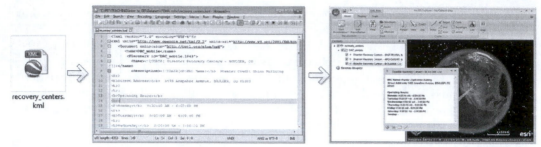

| KML is a text-based format | KML is form of XML. The XML tags define the structure and content of KML | KML can then be rendered in a variety of software |

Figure 3.7 An example of the KML format. In this example, FEMA disaster recovery centers are shown. (ArcExplorer screen shot Copyright © 2014 Esri, ArcGIS, ArcMap. All rights reserved. Used with permission.)

A proprietary spatial data format created by Esri in 1997, a *shapefile* is actually a collection of three or more files for storing vector GIS data (discussed later in this chapter) and has been a de facto (although not official) spatial data standard for many years due to Esri's large GIS market share and publication of the shapefile format (Environmental Systems Research Institute, 1998). Although Esri is deemphasizing shapefile data format in favor of their geodatabase file format, it is still a widely used format and many GIS datasets published by government entities in the United States, such as US Census Bureau TIGER files, and is thus important to mention. Shapefiles also store geographic information in a matrix (i.e., rows and columns) format, but raw shapefile data can only be viewed using special software unlike CSV files, which can be viewed using a basic text file viewing program like Notepad++ (http://notepad-plus-plus.org/).

Figure 3.7 shows the Keyhole Markup Language (KML) data storage format.

Originally developed by Google for use in Google Earth, KML has seen increasing popularity in recent years due to its ease of creation using Google Earth and the fact that KML is now an Open Geospatial Consortium (OGC) data standard (Open Geospatial Consortium, 2014). KML is an eXtensible Markup Language (XML) format and is thus an ASCII-based file format viewable in a text editor, making it useful for GIS applications that can read XML. As seen in Figure 3.7, KML contains both raw geographic information such as coordinates, but also the *presentation* of the geographic information such as colors used to display map features, text to display when a feature is clicked, and a wide variety of items such as time series and three-dimensional database solutions also exist.

OGC and Open Data Standards

The Open Geospatial Consortium (OGC; http://www.opengeospatial.org/) is an international standards body that maintains numerous geospatial and locational standards for a wide variety of application domains and industries. For example, the Web Map Services (WMS; http://www.opengeospatial.org/standards/wms) standard allows different GIS applications to share georegistered images with one another using a simple HTTP protocol. The Geography Markup Language (GML; http://www.open-geospatial.org/standards/gml) standard defines an XML-based grammar to define geographic features and is often used in streaming geographic data services. OGC standards are particularly important for disaster management applications as the standards can ideally allow for greater discovery, sharing, and interoperability of geographic information across different GIS technology platforms, data formats, and organizations. For more information on OGC disaster management activities, see the OGC website at http://www.opengeospatial.org/domain/eranddm (Figure 3.8).

Figure 3.9 shows imagery files.

Imagery, such as satellite or aerial imagery, is commonly used as reference data in GIS (refer back to Chapter 2 for discussion of reference maps and data). However, during disaster response, rapid image acquisition from a disaster zone can become critical to situation awareness and understanding and quantifying damage impacts (van Aardt et al., 2011). Imagery can be considered a form of *raster-type* geographic information, which is discussed later in this chapter. Imagery can be stored in a variety of file formats such as.tiff, geotiff, .jpg, .sid, and others. Some image files can be viewed in standard image software such as Adobe Photoshop. However, to view the geographically referenced images requires GIS software or other software designed for handling

Figure 3.8 The OGC Emergency Response and Disaster Management webpage. (From OGC. Used with permission.)

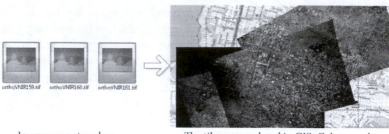

Imagery are stored as
individual files or "tiles"

The tiles are rendered in GIS. Colors can be
manipulated such as the black background
not shown in two of these tiles

Figure 3.9 A sample of imagery files collected after the 2010 earthquake in Haiti. These images were very valuable for viewing damage done to buildings in Port-au-Prince. (Imagery files from Rochester Institute of Technology, http://ipler.cis.rit.edu/projects/haiti; base map data behind the images, © OpenStreetMap contributors.)

An open source, relational database showing a table
definition for geographic features

Contents of the table can be rendered in GIS
tools that can connect to the database

Figure 3.10 Storing GIS data in a relational database. In this example, the open-source database postegreSQL/postGIS (left side of the figure) show geographic content in QGIS (right side of the figure).

geographically referenced imagery or other raster data such as Erdas Imagine or ENVI. Rather than handling large repositories of image files, images are increasingly being disseminated via web services.

Figure 3.10 shows storage of GIS data in a relational database.

The relational database is a very broad category of GIS data storage and entails keeping geographically referenced information inside structures that normally store other types of nongeographic information. Storage of GIS data inside of relational databases is typically used in large-scale operations where there are large volumes of GIS data that have complex modeling requirements and need to be shared with many people. Most professional-grade GIS technology offers support for storing GIS data inside relational databases such as Microsoft SQL Server. However, as seen in Figure 3.10, open-source GIS enterprise database environments are also available.

Analysis

Analysis refers to the use of GIS to investigate geographically or spatially oriented questions or problems. An important point in this regard is that GIS software contains methods or tools designed to assist in understanding spatial patterns or processes. Disaster management lends itself well to providing specific examples of GIS analysis due to the fundamentally spatial nature of disasters.

As a hypothetical, yet real, GIS analysis example, based on events that happened in 2012 Hurricane Sandy—a disaster manager may wish to understand how to reduce the risk of people who are vulnerable to a flood hazard (Parry, 2013). She hypothesizes that a large storm swell will likely affect a larger number of elderly people than has been currently accounted for and are unlikely to seek shelter (Saul, 2012). To test her hypothesis, she starts by bringing census data into her GIS that indicates how many elderly people are living in a flood zone and next to a shoreline. She then uses a buffer tool to calculate distances from the shoreline to spatially understand how many elderly people may be affected by different storm surge extents (Figure 3.11).

GIS Programming

GIS programming refers to the use of computer programming languages to build custom software applications or tools to accomplish tasks that out-of-the-box GIS software might not be able to accomplish. In the early days of GIS, operations in GIS software were all

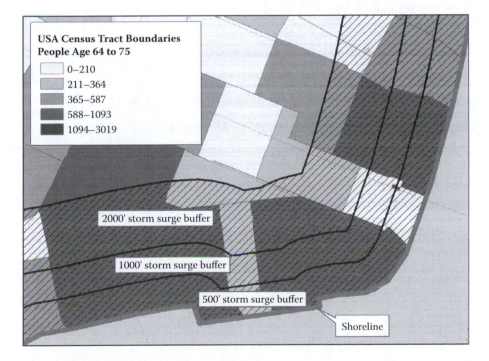

Figure 3.11 Buffer example; calculating distances from a shoreline to understand potential impacts on vulnerable populations.

conducted using programming languages such as FORTRAN and AML (ArcMacro Language) to send commands to the software. With the rise of Windows and personal computers in the 1990s, GIS software evolved to allow operations and interactions based on GUIs and the mouse. GIS programming thus evolved to become a more specialized task requiring interdisciplinary computing and information technology knowledge and skills such as computer programming that could be matched with GIS software tasks and principals. GIS programming is still a highly valued skill, knowledge of which makes one very valuable in terms of employability. A GIS programmer may write computer code for tasks ranging from batch data processing and automation to the development of modern-day *mapping mashups* that use complex algorithms for integrating heterogeneous data sources to solve unique problems (Batty et al., 2010; Liu and Palen, 2010). At the time of this writing, important computer programming languages to know for GIS programming are JavaScript for development of web-based GIS applications, Python for scripting tasks inside of major commercial GIS packages such as Esri's ArcGIS, and languages such as Java, C# or .Net for development of desktop GIS applications or native mobile device applications.

Mapping APIs

GIS programming is often based on the use of mapping application programming interfaces or APIs. Mapping APIs allow computer code to be written that utilizes objects, methods, and functions with the APIs. The following block of JavaScript code from the popular OpenLayers API demonstrates these ideas (Figure 3.12).

Many mapping APIs exist that can be used to support GIS programming tasks based on the underlying technology that will be programmed. A nonexhaustive collection of mapping APIs for web environments as of 2014 include

Google Maps API for web, phone, and tablet environments: https://developers. google.com/maps/

Esri APIs for JavaScript, Flex, and Silverlight development platforms: http:// www.esri.com/software/arcgis/apis

Microsoft Bing Maps API: http://www.microsoft.com/maps/choose-your-bing-maps-api.aspx

Like any specific technology, make sure to review any updates to these technologies and their corresponding URLs since the printing of this book.

Modeling

Much like model trains or cars give us a scaled representation of a real-world entity, modeling in the context of GIS is the idea of using GIS to simulate conditions in the real world to answer what-if questions. For example, a GIS-based model could be developed to simulate possible storm surge conditions and outcome scenarios. Furthermore, a powerful analytical capability of GIS-based models is the ability to tweak parameters

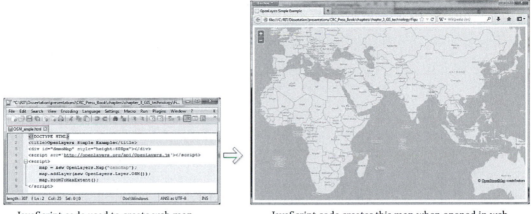

JavaScript code used to create web map. Note line 4 which connects the code to the OpenLayers API

JavaScript code creates this map when opened in web browser.

Figure 3.12 Using the OpenLayers API to create a simple web map using OpenStreetMap data. (Code based on example from http://wiki.openstreetmap.org/wiki/OpenLayers_Simple_Example.)

within a model to evaluate different conditions and test relationships between different model parameters (Maguire, Batty, and Goodchild, 2005). Using the storm surge example, parameters that might be modified and observed to evaluate different scenarios including the strength of the storm surge and the time of day when the storm surge is happening to understand population impacts. For example, a model with a 3-meter storm surge within a city at 11:00 a.m. on Wednesday will produce vastly different results than the same model run on Saturday at 10:00 a.m. due to the effects of differing populations with a city between weekday and weekend work schedules of people. As we will see in later chapters, GIS-based models can be developed for almost any aspect of the disaster management cycle.

GIS-Based Disaster Modeling Tools

Several GIS-based modeling tools have been developed by US government agencies. Three important spatial modeling environments for disaster management are as follows:

HAZUS: This model focuses on estimating loss from natural hazards, specifically earthquakes, floods, and hurricanes. Through GIS-based modeling, HAZUS can estimate social, physical, and economic disaster impacts. Visual representations are a key component of HAZUS as the power of maps are utilized to show spatial relationships between a natural hazard and items such as populations or other resources (see http://www.fema.gov/hazus for more details).

Areal Location of Hazardous Atmospheres (ALOHA): This model is designed to model the spatial distribution of hazardous gases. ALOHA can model parameters relevant to the release of toxic, gas-based substances such as the pressure of the container holding the gas, the size of the opening in the gas container, gas storage temperature, wind speed and parameters to determine exactly what type of gas plume would be generated from the release. The gas plume that ALOHA generates can then be overlaid in a GIS to answer what-if questions about the plume such as, "What if the plume spreads over a residential area when people are home?" (Figure 3.13).

Standard Unified Modeling, Mapping, and Integration Toolkit (SUMMIT): SUMMIT is a modeling environment supported by the US Department of Homeland Security "that enables analysts, emergency planners, responders, and decision makers to seamlessly access integrated suites of modeling tools & data sources for planning, exercise, or operational response" (Standard Unified Modeling Mapping, and Integration Toolkit, n.d.) (Figure 3.14).

ALOHA generates plume model

Plume model added to a GIS to answer "what if" questions about toxic gas release in a city

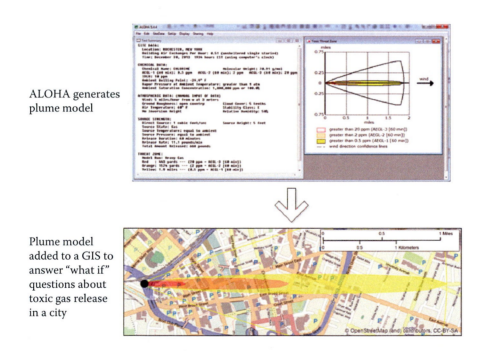

Figure 3.13 A plume generated in ALOHA (top of figure) that is then imported as a KML file into a GIS and layered on a map to show how the plume would affect the area of interest. ALOHA is US government software. For more on using ALOHA with GIS, see http://response.restoration.noaa.gov/aloha and Tomaszewski, Brian. 2003. Emergency response and planning application performs plume modeling. *ArcUser*, 10–12. (Figure based on map data © OpenStreetMap contributors.)

Figure 3.14 SUMMIT screenshot. For more information, visit: https://dhs-summit.us/.

Cartography, Visualization, and Map Production

As you saw in Chapter 2, important GIS functions are map production and cartography support tools. In essence, map production can be seen as the final process related to the other functions previously discussed in this chapter. For example, a map can be used to represent the final results of a GIS analysis to give to a decision maker or be used to represent different parameters, scenarios, and outcomes from GIS-based modeling to make modeling results easier to understand. Commercial desktop GIS tools such as Esri's ArcMap come with comprehensive tool sets to support the processes behind the art and science of cartography as well as numerous tools for final map product outputs for printing, use as static digital graphics, or map tiles within dynamic, web-based reference maps. The rise of online mapping tools like Google Maps is also changing long-held conventions about cartography and map production as these types of technologies in some way limit the cartographic process as the map maker is restricted, for example, to only using the Google base map as a reference map, which may not always be the best choice (Field and O'Brien, 2010). However, arguments for using mapping tools like Google Maps and Google Earth for cartographic and map production in disaster management cases, especially disaster response, are compelling when one considers the speed and ease of use by which these tools can produce maps. For example, quick situation awareness maps can be made by plotting point features on top of Google Earth's visible satellite imagery maps (Figure 3.15).

Figure 3.15 Viewing a USGS Earthquake KML feed inside Google Earth.

Geocoding

Geocoding is the idea of taking text-based input such as a place name or street address and converting it to a coordinate representation. For example, "1600 Pennsylvania Avenue, Washington, DC" would be geocoded to 38.897881, −77.036530 in decimal degree coordinates. A common, everyday example of geocoding is entering the name of a place, business, or address in a tool like Google Maps that can then quickly geocode the item entered and show its location on a map. Geocoding has numerous important uses for disaster management that include, but are not limited to, geocoding (or more accurately, geotagging) picture locations, geocoding tax parcels in relation to flood zones, and address searching for missing people (Schradin, 2013).

Limitations of GIS

Although the overall intent of this book is to inspire and help you learn about GIS for disaster management, it is important to also consider the limitations of GIS in disaster management. Technology in general is often seen as a miracle cure for existing problems, but it is important to manage the expectations about what GIS can do. The following are some points to keep in mind in terms of the limitations of GIS.

GIS software is not a miracle technology that can automatically answer all questions. Although this may seem obvious, it is important to keep in mind that GIS is limited by the numerous components of the system that comprises GIS, as discussed previously. For example, the answers you get are only as good as the software used, the quality of the data used in the software, and the skills of the people operating the software, conducting the analysis, and modeling and producing the final maps. GIS can strongly support answering questions, but it is still human reasoning and critical thinking that must make final decisions. Overreliance and overexpectation of the technology coupled with lack of proper GIS education and training and lack of good human judgment, reasoning, and critical thinking, can all lead to dire consequences.

The acquisition, creation, editing, and duration of data is the most costly aspect of GIS. Anyone who is experienced with GIS has most likely learned this lesson the hard way. If you are new to GIS, it is very important to understand the importance of data for being successful at utilizing GIS technology for disaster management. GIS operations, analysis, modeling, and cartography are fundamentally data driven. Acquiring GIS-ready data is both financially costly in terms of hours spent collecting and editing data, or perhaps spending money on purchasing GIS data from a data vendor such as Navteq (http://www.navteq.com/). In my own teaching experiences, I have seen many great student research project ideas fail or have to undergo major modifications due to lack of data to support the investigation. Thus, if you are new to GIS, pay close attention to how you will find data that can support your investigation and how much time and possibly money you are willing to spend to acquire data. In the disaster management context, data handling must be done during planning (discussed in Chapter 5), because when it comes time for a disaster response, there is no time to acquire data and spatial deluge may occur (discussed in Chapter 6). At the end of this chapter, a list is provided of free GIS data sources relevant to disaster management.

The following sections discuss GIS data models and specific GIS software technology.

UNDERSTANDING GIS DATA MODELS

An important concept to understand when working with GIS data are GIS data models. A GIS data model can be thought of as a way in which geographic-scale features or phenomena are represented in a digital manner. Remember that digital representations ultimately reduce the item being represented to binary 1s and 0s. Reducing geographic reality to 1s and 0s is problematic, and many of the nuances, subtleties, and idiosyncrasies are lost in digital representations—a problem that really has existed since the beginning of mapping. For example, in a disaster management context, how does one represent the shifting nature of an eroding shoreline or differences in vulnerability that do not lend themselves to simple polygon-based representation. Thus, GIS data models have been developed to address these representational issues in a variety of manners (and varying degrees of success). The two most common forms of data GIS models are *vector* and *raster*.

Vector Models

The vector GIS data models represent geographic features as discrete, vertex-based shapes. Each vertex in a vector shape is referenced to a specific x,y Cartesian or geographic coordinate location. By using a discrete, vertex-based approach, the vector data model is generally advantageous for representing geographic features that have *discrete* boundaries or edges or can be reduced to representation as a single x,y coordinate pair. In practical terms, this means that the vector GIS data model is typically used to represent geographic features as points, lines, and polygons as these geometric primitives intuitively lend themselves to representation of features or phenomena that have discrete edges or boundaries. Figure 3.16 graphically represents a disaster management example of vector-based points, lines, and polygons to illustrate these ideas.

An important aspect of vector-based GIS datasets from a technical perspective is that for any given geographic feature stored in a vector dataset, nonspatial attribute data can be stored along with the data that describes the vertices of the point, line, or polygon shape itself. Figure 3.17 demonstrates this idea.

Having nonspatial attributes associated with geographic features, as per the vector GIS data model, is one of the fundamental analytical features of GIS. For example, thematic maps can be based on qualitative or quantitative attributes (as discussed and shown in Chapter 2) or nonspatial attributes can be queried using Structured Query Language (SQL) statements to ask questions of GIS data (Figure 3.18).

In most GIS software packages, the actual specific data that defines the point, line, or polygon contents are hidden from the end user and the GIS software itself takes care of editing the vertices. However, the increased use of text-based, XML-structured GIS data formats such as KML or GeoJSON demonstrate how vector shape vertices coordinates are in fact human-readable. Having vector shape vertices coordinates in human-readable, text-based format is significant because GIS software-readable data can, in many cases, be created and edited without specialized GIS tools (Figure 3.19).

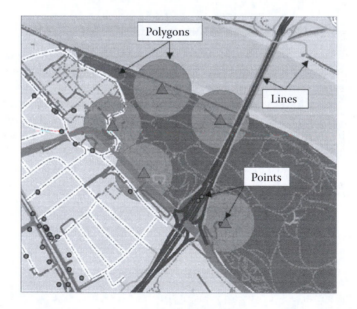

Figure 3.16 An example of vector points, lines, and polygons in a disaster management context. In this hypothetical scenario, toxic waste barrels have washed up on shore from a river and are presented as points using triangle symbols. A 500-foot buffer has been generated around the points to create polygons that show the toxicity threats of the barrels in relation to line features such as roads and bridges. (Map by Brian Tomaszewski.)

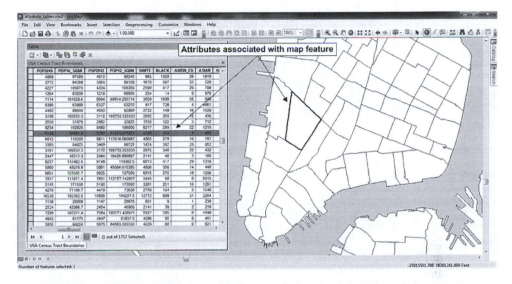

Figure 3.17 Attributes associated with a vector polygon. In this example using US Census tract boundaries, one record from the attribute table has been selected and the corresponding map feature associated with this record is outlined in black. Make note of how the attribute table contains a variety of different attributes that can be associated with a feature such as population counts or racial composition. (ArcMap screenshot Copyright © 2014 Esri, ArcGIS, ArcMap. All rights reserved. Used with permission.)

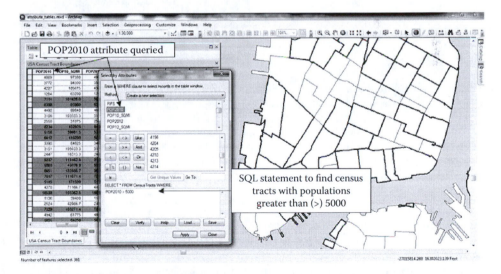

Figure 3.18 Querying attribute tables with SQL statements to answer questions. In this example, a query was made to find census tracts whose population in 2010 was greater than 5,000 people (as specified in the POP2010 column). The left side shows how 361 features matched this query criteria and the map on the right side correspondingly shows a variety of selected features. (ArcMap screenshot Copyright © 2014 Esri, ArcGIS, ArcMap. All rights reserved. Used with permission.)

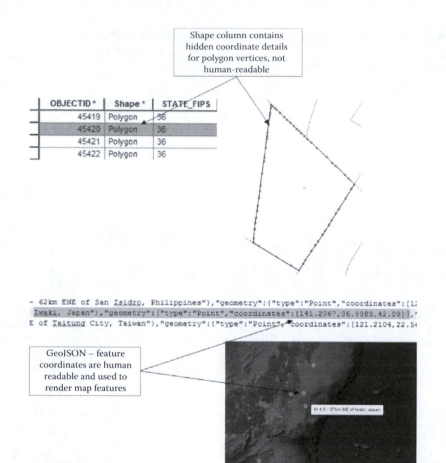

Figure 3.19 Coordinate information in vector features. In the top-right part of this figure, a polygon feature and the vertices that define its shape are shown. Note how the attribute table has a shape column that only indicates "polygon" and does not display any vertices coordinate details. The middle of this figure shows an excerpt of earthquake location GeoJSON code. Note how the coordinate values are human-readable, and thus potentially editable with a text editor, and can be used to create vector point features like those shown on the map in the bottom right of this figure.

Raster

The raster GIS data model represents geographic features and phenomena as a grid of individual cells. As opposed to vector, raster is typically used for modeling geographic entities that are *continuous* in nature and have no discreet boundaries or edges. Typical examples of continuous phenomena that would be represented in the raster data model include temperature and elevation. Raster data is also the format used for imagery ranging from aerial photography to space-based images that can be can be incorporated into GIS software.

Figure 3.20 provides specific disaster management examples of raster GIS data.

Figure 3.20 Disaster management raster data examples. The top part of this figure shows a digital elevation model (DEM) captured during the 2010 Haiti earthquake. Note how the grid of cells shows the varying landscape elevation—a representation technique that is difficult to achieve with vector data. In the bottom portion of of the figure, two raster datasets are shown side by side. On the bottom figure, left side, a DEM is shown; note how the individual cells can be seen. On the bottom figure, right side, an image of the presidential palace is shown. For imagery, each pixel in the raster grid represents a color in the image. DEMs and imagery are often combined in disaster management to understand how the underlying landscape interacts with the built environment, for example, the relationship between water runoff from slopes and building locations. Also differing from the vector data model, the raster data model stores numerical values associated within each grid cell; for example, the elevation of a given cell or its temperature.

An important concept with raster data is *spatial resolution*. Spatial resolution is fineness of detail of a raster dataset and is based on the size of each cell within the grid. For example, the smaller each grid cell, the finer the spatial resolution of the raster dataset. This idea is no different than that of digital camera picture pixel resolution. Spatial resolution is an important concept to be aware of with raster data, much like the discussion of map scale in Chapter 2. Spatial resolution is an important consideration for determining the appropriateness of a raster dataset for a given purpose. For example, too coarse a spatial resolution may not provide enough detail for a given task, while too fine a resolution may not provide enough coverage of a given area. Figure 3.21 provides examples of specific raster datasets to illustrate these points.

Now that you have learned about GIS data model basics, the next important concept to understand is GIS metadata.

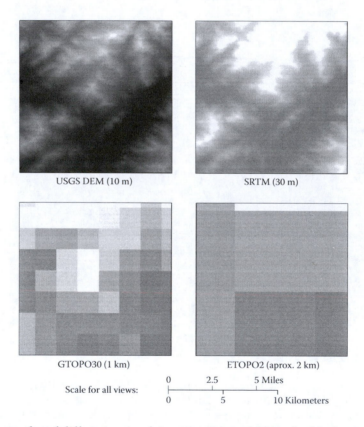

Figure 3.21 Examples of different raster dataset spatial resolutions. In this figure, the same overall geographic extent is shown in each view (as per the scale bar shown in middle). However, each view shows how spatial resolution varies. For example, the top left view shows the USGS DEM at 10-meter spatial resolution, the top right view shows data from the Shuttle Radar Topography Mission (SRTM) at 30-meter spatial resolution, the bottom-left view shows data from the Global 30 Arc-Second Elevation (GTOPO30) at 1-kilometer spatial resolution and the bottom-right view shows data from the Global Digital Elevation Model (ETOPO2) at approximately 2-kilometer spatial resolution. Note the striking contrast seen when using different raster spatial resolutions.

GIS METADATA

Metadata is *data about data*, or more precisely, data records that describe the contents, characteristics, lineage, or anything else about another dataset. For example, when you go grocery shopping, you typically do not buy canned items that do not have labels on them, as you would have no idea what is inside of the can. Metadata is the label on the can. This same idea works with digital GIS datasets whether they are raster or vector. For example, metadata for a vector GIS dataset might describe the geometry type used (point, line, or polygon), the attributes found in each record, and who initially created the dataset. Metadata for a raster dataset might describe the spatial resolution of the individual cells and the geographic extent of the overall grid, as well as the specific data format used for storing numerical information in each cell such as integer or floating point numbers. The structure and physical storage of GIS metadata is as varied as GIS data itself. For example, in the United States, the US federal government Federal Geographic Data Committee (FGDC; http://www.fgdc.gov/) has a specific metadata standard to which all GIS datasets created by the US federal government agencies must adhere. FGDC-compliant metadata must contain descriptions such as the following:

1. *Identification Information:* This identifies who created the data.
2. *Data Quality Information:* This information concerns how the data was created and any quality control issues encountered during the creation such digitization errors and the data lineage (i.e., who has worked on the data).
3. *Spatial Data Organization Information:* If the data is being organized under some type of formal organizing structure such as the Federal Information Processing Standards (FIPS) used in the united States.
4. *Spatial Reference Information:* Spatial reference Identifies the coordinate system and geodetic model (i.e., datum) used.
5. *Entity and Attribute Information:* Entity and attributes include descriptions of what attributes are used in the dataset and what the values of the attributes mean (in cases where a code is used instead of an actual value).
6. *Distribution Information:* This information tells how the data can be distributed and whom to contact to obtain a copy of the data.
7. *Metadata Reference Information:* This identifies the standard the metadata itself is using.

Figure 3.22 is a screen shot of a metadata file in HTML format containing each of these description categories.

Metadata plays a particularly important role in GIS for disaster management. This importance is due to the fact that metadata is critical to coordination and collaboration activities, as discussed in Chapter 1. For example, increases in larger disasters that span multiple jurisdictional or even national boundaries make it vitally important that disaster managers know what data they have to work with and how appropriate a given dataset is to a task at hand. In time-sensitive situations, there is no time to evaluate the usefulness or fit of a dataset. Proper, updated documentation of GIS datasets through metadata is vital in making informed decisions about GIS data.

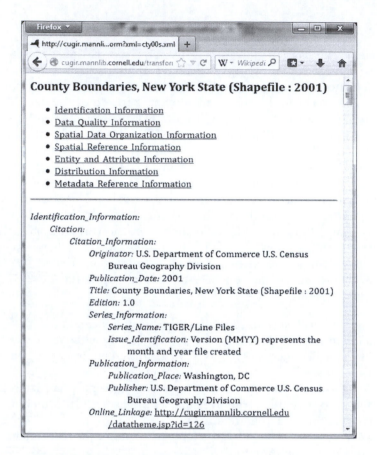

Figure 3.22   Example of HTML-based GIS metadata.

SPECIFIC GIS TECHNOLOGY

Now that you have a solid background in GIS concepts, the following section presents specific GIS technology. This is perhaps the one part of the book where you will need to be most careful in checking that specific items mentioned are still in existence and available, and that URLs listed here have not changed. Even though all technology changes, every effort has been made to ensure that the technology discussed in this section will likely be available for at least five years after this book is published. More specifically, the criteria used for the specific GIS technology in this section include

- the technology has been around for at least 10 years prior to publication of this book;
- the technology is being actively supported through ongoing maintenance and support by a corporation that sells it or an actively supported open-source software development community; and
- the technology has documented use in disaster management practice.

Please note that are numerous types of GIS technology applicable to disaster management and that many volumes could be published on the many technologies that exist. The technologies listed here are meant as a guide to the general varieties that exist, and you are encouraged to do your own research or create your own technologies to fit your GIS for disaster management needs.

GIS Technology Platforms and Disaster Management

In today's ever-increasing interconnected world, GIS is available for all types of computing platforms ranging from traditional desktop systems, to phones and tablets, to virtual, cloud-based environments that provide GIS functionality through web interfaces. It is important, however, to distinguish which technology platform is appropriate for which aspect of disaster management work. For example, developing a complex flood hazard model will be very difficult to accomplish on an iPhone due to limited computing resources and screen space. Conversely, viewing simple point locations where relief stations are located does not require a supercomputer cloud cluster. Thus, GIS is evolving to where different technology platforms are being used for specific tasks and needs of specific consumers of GIS data, services, and products, such as maps. For example, desktop computers are still best suited for core analytical and data management tasks for disaster management due to the heavier computing power that is available. In the past ten years, with the decrease in demand for specialized desktop GIS software and the increased demand for more lightweight tools focused on the viewing of GIS and map-based data for things like location-based services, GIS technologists have begun to offer GIS technology on what are known as "thinner" (i.e., less CPU/RAM and overall computing power) clients such as mobile and web-based platforms. For disaster management, mobile platforms are increasingly showing their benefit for allowing disaster management practitioners to access GIS data contained within larger systems and to collect data from the field using simple point collection procedures that utilize the GPS receiver common to most mobile platforms.

An advantage of web platforms is that they can be accessed anywhere that there is an Internet connection; they are (usually) not restricted by operating systems, plug-ins, or choice of web browser if they conform to various W3C (World Wide Web Consortium) standards for coding HTML pages. A very interesting development at the time of this book's writing is the advent of HTML5. HTML5 is allowing for more interaction and functionality in web browsers. In the past, web browsers typically required the use of a plug-in, such as the Java or Flash Player, which was often a barrier to web application use. For example, with HTML5, an application can be designed so that it can run either in a web browser or on a mobile device, and all from a single code base.

In the following sections, specific GIS technologies are discussed from the dual perspectives of technology platform and disaster management tasks and end users.

ArcGIS

ArcGIS is an umbrella term for a wide range of GIS technologies created and maintained by Esri, the world's largest commercial GIS software company. Esri GIS technology is used in countless disaster management organizations around the world as evidenced by the numerous disaster management and homeland security success stories Esri maintains (Environmental Systems Research Institute, n.d.; Kataoka,2007). Esri offers a comprehensive range of GIS technology relevant to all modern technology platforms and disaster management tasks and end users.

For *desktop* applications, there is the ArcGIS/ArcMap (http://www.esri.com/software/arcgis) application that has numerous, powerful features. Select examples of features with strong relevance to disaster management are illustrated in the following series of screen shots (Figures 3.23–3.25) taken from ArcMap:

The very rich feature set and overall complexity of the software make ArcMap a somewhat challenging tool to use. Often, those who are most competent in its use have received specialized training and education.

For *mobile platforms* such as Android and iPhone-based operating systems, Esri offers a variety of APIs for software developers to build custom GIS applications. Furthermore, Esri offers several free apps that the public can download view data and perform some analytical procedures such as map overlay and changing map displays.

For web platforms, Esri also offers several APIs that software developers can use to build custom web-based GIS applications (as discussed previously). At the time of this writing, APIs for JavaScript, Flex, and Silverlight web development environments are offered, but be sure to check which ones are still supported if you are interested in examining web-based technology from Esri.

In recent years, Esri has strongly emphasized its ArcGIS Online technology. ArcGIS Online is a paid, subscription-based service that provides access to many functionalities offered through desktop ArcMap, but in a web environment. Furthermore, ArcGIS Online

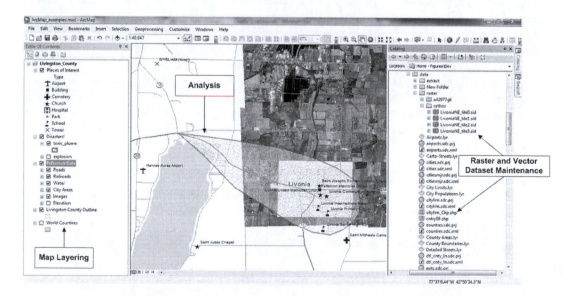

Figure 3.23 A disaster management example of map layering, vector and raster dataset maintenance, and analysis in ArcMap. In this example, impacts from a toxic plume cloud are being investigated. (Screenshot Copyright © 2014 Esri, ArcGIS, ArcMap. All rights reserved. Used with permission.)

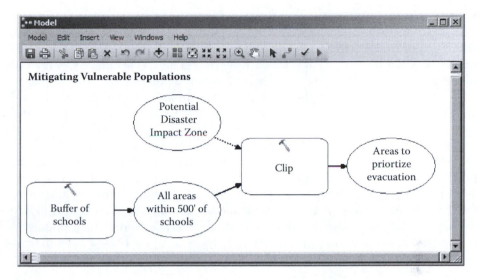

Figure 3.24 A disaster management example of modeling tools available in the ArcMap model-builder tool. ArcMap models allow a series of GIS processes and datasets to be combined in one workflow. In this simple example, a model is used to determine areas to prioritize for evacuation. (Screenshot Copyright © 2014 Esri, ArcGIS, ArcMap. All rights reserved. Used with permission.)

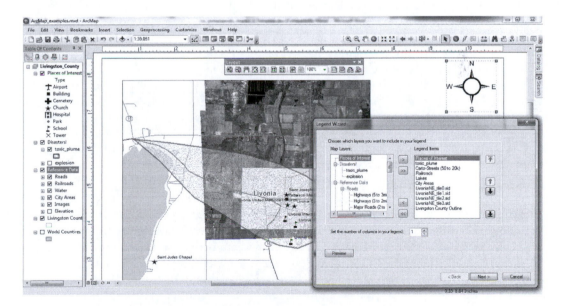

Figure 3.25 A disaster management example of select cartographic production tools available in ArcMap. ArcMap's map layout functions make it easy to produce maps using tools such as a legend wizard and automated scale bar and north arrow generation. (Screenshot Copyright © 2014 Esri, ArcGIS, ArcMap. All rights reserved. Used with permission.)

offers an impression range of premade reference maps, apps, and analytical tools that allow users to quickly start using GIS functionality and to interact and share data with one another within the ArcGIS Online community. ArcGIS Online is also a cloud-based environment, meaning that its capacities are very scalable and available wherever there is an Internet connection and across all types of hardware platforms.

In disaster management applications, ArcGIS Online is seeing increased use for the sharing and display of disaster-related situational information. For example, the Federal Emergency Management Agency (FEMA) GeoPlatform (http://fema.maps.arcgis.com/home/; discussed further in Chapter 4) uses ArcGIS Online to display a wide range of GIS data from FEMA projects such as cyclone impacts, flood hazards, risk assessment, and more. Esri's increased emphasis on their ArcGIS Online technology is part of a growing trend in cloud-based, web-driven applications that provide GIS analytical functionality, maps, and datasets to end users for a wide variety of disaster management contexts ranging from simple maps that display point locations to complex analytical prediction models. Perhaps the most visible and widely recognized company, outside of the traditional GIS world, that has been developing web-driven technology for disaster management is Google.

Google Maps and Other Google Geospatial Technology

As an Internet search company that "grew up" on the web, Google has a well-established record of developing web-driven applications that work with the massive datasets that comprise the web. Their flagship geospatial technology, Google Maps, is perhaps the most recognized mapping technology in the world. For example, when I teach introductory geospatial technology students, at the start of each semester, few to none of have heard of Esri, but all of them are familiar with Google and many report using Google Maps almost every day. Of course, Google's geospatial technology goes far beyond Google Maps to include the virtual globe Google Earth, mapping APIs, location-oriented services, and even the entire Android operating system, the world's most widely used mobile device operating system, which contains native location functionality for GPS receiver access and more. Google has also made their mapping tools more accessible to nontechnical experts through technology such as Google Maps Engine (https://mapsengine.google.com), which allows end users to create content on top of the Google base map without having to use computer programming languages such as JavaScript to do so.

Google is also consistently part of major worldwide disaster response and relief efforts. Specifically, Google's philanthropic division (google.org) has Google Crisis Response (http://www.google.org/crisisresponse/), a team dedicated to using the Internet to collect data and build tools during major disasters. A very common approach that is used by the Google Crisis team for disseminating crisis data is to host a *crisis response map* that displays thematic data related to a given crisis inside a Google Map interface (Figure 3.26). Furthermore, Google makes data for a given crisis available in the KML format, which allows for the easy sharing of geographic data across multiple GIS platforms such as Google Earth, but other technology as well, such as ArcMap.

Google technologies continue to see increased use for disaster management applications due to the ubiquitous nature of Google technology in general, ease of use, familiarity that people have with using Google technology, and the fact that there is (usually)

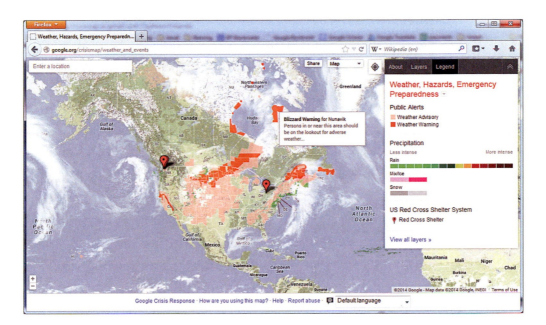

Figure 3.26 Google crisis map.

no cost involved in using tools like Google Maps and Google Earth. However, other *free and open-source* GIS technology solutions exist, many of which provide a better technology solution for disaster management practice based on limited financial resources, unreliable Internet connectivity, or a general desire to avoid commitment to a particular software company that ultimately has control over critical technological assets. In the open-source GIS world, numerous options exist; the following are a few of the most widely used.

QGIS

QGIS (http://www.qgis.org/en/site/) is perhaps the most widely known and used open-source GIS package. QGIS offers many of the same features as found in commercial GIS tools such as ArcMap (Figure 3.27).

For example, QGIS can handle raster and vector data, map production, and offers a range of spatial analysis tools and modeling capabilities. Additionally, QGIS offers a Python-based scripting environment that allows for custom scripting and specialized GIS tools to be developed inside the QGIS environment. QGIS is also capable of working with other popular open-source GIS tools such as Geographic Resources Analysis Support System (GRASS) (http://grass.osgeo.org/) and the Geospatial Data Abstraction Library (GDAL) (http://www.gdal.org/).

In terms of QGIS and disaster management activities, QGIS is commonly used around the world for disaster management training, and an Internet search on the terms "QGIS disaster management" will reveal numerous training opportunities and information on open-source GIS disaster management solutions that feature QGIS.

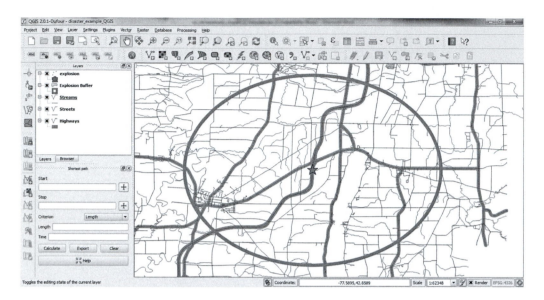

Figure 3.27 A disaster management example in QGIS. In this example, an explosion has occurred and the extent of the explosion has been buffered. Note how QGIS has a similar look to ArcMap and offers many of the same functionalities. QGIS is covered under Creative Commons Attribution-ShareAlike 3.0 license (CC BY-SA) http://creativecommons.org/licenses/by-sa/3.0/.

Other Commercial, Free, and Open-Source or Openly Available GIS Technologies

The GIS technologies that were discussed so far in this chapter represent some the most widely used and well-known GIS technologies. However, they are not the only available options. The following sections outline a variety of other GIS technologies that are used in disaster management.

OpenStreetMap

As discussed in Chapter 2, OpenStreetMap (OSM) provides a platform for users to create and edit a freely editable world map. The data provided through OSM continues to serve a vital role in disaster management activities for areas where reference data is not available. OSM and the Humanitarian OpenStreetMap Team (HOT; http://hot.openstreetmap.org/) will be revisited once again in Chapter 4, which discusses various organizations involved in disaster management.

Other GIS Technologies

The following general GIS technologies could also be used for or could support a variety of disaster management activities:

- *OpenLayers* (http://openlayers.org/) is an open-source, web-based map client similar to the Google Maps API.

- *Mapbox* (https://www.mapbox.com/) is a free commercial map service that utilizes OpenStreetMap data and places particular emphasis on cartography and design of maps created in the MapBox system.
- *MapServer* (http://mapserver.org/) is a free and open-source publishing service that allows map data to be served over the web.
- GeoServer (http://geoserver.org/display/GEOS/Welcome) is similar to MapServer in that it is a free and open-source service to publish map-based data over the web.
- *PostGIS* (http://postgis.net/) is an open-source spatial database that works with postgreSQL (http://www.postgresql.org/) to provide geographic objects and spatial queries
- *NASA World Wind* (http://worldwind.arc.nasa.gov/java/) is an open-source virtual globe tool that is particularly suited for viewing datasets created by NASA, such as the Landsat series.
- *MapInfo* (http://www.mapinfo.com/) is commercial GIS technology by Pitney Bowes, and is similar to Esri technology in that it offers desktop, web, and server-based GIS tools.

Free and Open-Source Datasets Relevant to Disaster Management

The following is a nonexhaustive list of reference and thematic sources of GIS datasets that are relevant to disaster management and can be downloaded and used in the technologies previously mentioned:

- *The National Map* (http://nationalmap.gov/) provides access to numerous reference and thematic data layers for the United States.
- *USGS Global Visualization Viewer* (http://glovis.usgs.gov/) provides free access to numerous NASA and USGS products such as LANDSAT, MODIS, ASTER, and TERRA.
- *Global Administrative Areas* (http://www.gadm.org/) provides free access to worldwide administrative boundaries.
- *GeoNames* (http://www.geonames.org/) provides free worldwide gazetteer (place name) data; a good source for geocoding applications.
- *FEMA GIS Data Feeds* (http://gis.fema.gov/DataFeeds.html) provides a collection of FEMA's disaster declarations and support offices relative to the United States.
- *US Census Bureau American FactFinder* (http://factfinder2.census.gov/faces/nav/jsf/pages/index.xhtml) is relevant to the United States, and can be used to find a wide variety of census indicator data (discussed again in Chapter 8).
- *US Census Bureau TIGER* (http://www.census.gov/geo/maps-data/data/tiger.html) is relevant to the United States, and provides a variety of vector features that can be used with American FactFinder data.
- *USGS Earthquake Hazards Program* (http://earthquake.usgs.gov/earthquakes/feed/v1.0/) provides a variety of feeds (including KML and GeoJSON) related to worldwide earthquake instances.
- *World Bank Data* (http://data.worldbank.org/) provides economic and social country-level indicators; data is in CSV, XML, and Excel formats.

- *United Nations Office for the Coordination of Humanitarian Affairs (UN-OCHA) Humanitarian Response Common Datasets* (https://www.humanitarianresponse. info/applications/data) provides "critical datasets that are used to support the work of humanitarian actors across multiple sectors. They are considered a de facto standard for the humanitarian community and should represent the best-available datasets for each theme." (quote from website)

Finding GIS Data on the Internet

As stated previously, finding or creating data is the most time-consuming aspect of GIS. A very important skill to learn is how to use the Internet to find GIS data. In the United States, most states have *GIS clearinghouses*, which (often) allow free access to public GIS data. At the US federal level, spatial data infrastructures (SDIs) such as GeoPlatform.gov provide access to a wide variety of GIS and other data produced by the federal government (Figure 3.28).

Depending on where you live, you may need to do some "hunting" on the Internet to find the right GIS datasets for your disaster management needs. Always be sure to look for metadata that describes the dataset and make sure that the GIS technology you are using can work with the dataset. For example, many datasets created in the old ArcINFO coverage format can still be found on the Internet, but will not work in a tool like Google Earth unless using special data conversion software.

Figure 3.28   The GeoPlatform.gov website.

How to Choose the Right GIS Technology for Disaster Management

There is no single guideline or set of principles to determine the right GIS technology to use for disaster management. There are several interconnected factors to consider and tie back to the components of GIS discussed in the beginning of this chapter.

First and foremost is cost; does your organization have the financial resources to afford purchasing and maintaining licenses with commercial software like MapInfo or Esri? (It is important to note that Esri does offer their software for free to nonprofit organizations; see http://www.esri.com/nonprofit.) If it does not, does your organization have the technical capacity to work with open-source software that often requires information technology–oriented knowledge of how to install and host open-source tools? Furthermore, how important is it that your organization have immediate technical support when technology problems arise? Commercial GIS software does give the advantage of a stable, reliable company that supports the products it creates and it can be quite reassuring to make an old-fashioned telephone call to get technical support versus posting a question to an online discussion forum for an open-source or openly available tool with no guarantee of getting a response (even in large communities like Google).

Second, what are the specific disaster management tasks you will have to conduct and how might those tasks change over time? For example, a small community disaster training organization might find Google Maps Engine or CrowdMap (discussed in Chapter 6) to be sufficient technologies for creating simple maps to show where disaster shelters are located for planning that can then can be used to monitor relief request locations during an actual disaster versus a large government organization that needs the computing power of robust GIS tools to develop flood hazard models that in turn have to be disseminated to millions of people via the web.

Third, how integral is GIS to your overall organizational mission? If GIS and maps only play a minor role in your overall activities, using robust and comprehensive GIS technology like desktop ArcGIS might not be the best choice. Technologies like these are difficult to learn and knowledge of how to operate them can be quickly lost if not maintained or the consultant or student intern hired to work with them has moved on from the organization. In the case of disaster response, the last thing that should be a factor in a response situation is trying to scramble around and learn (or relearn) how to use a technology that has been dormant for a long time. Besides these general items, there are myriad things to consider when choosing a GIS technology solution. Many GIS consulting firms provide GIS needs assessment consulting services that can help with making the right decisions; that is an option to consider if your organization has the resources to purse this route, or you can review sources such as Tomlinson (2007).

Getting Started with GIS Technology and GIS Technology Configuration Ideas

The sheer number of GIS technologies available coupled with the numerous tasks for which GIS can be applied in disaster management can make the process of getting started with GIS for disaster management a daunting task. The following section is written for those that are brand new to GIS or are looking to expand their use of GIS technology for disaster management beyond simple point mapping on Google Maps. Please note that step-by-step instructions and training on specific GIS technology is not within the scope of this book. For detailed steps on how to use a specific GIS technology, consult the help guides and tutorials that often accompany the software in question. If you find a technology that is not well documented or does not offer tutorials, that might help you determine if the technology is worth your time.

If you are brand new to GIS (i.e., you have never taken a class on the topic and have done minimal to no web searching on GIS), I recommend you first start with a free tool like ArcGIS Explorer (http://www.esri.com/software/arcgis/explorer). ArcGIS Explorer will give you a sense of map layering and working with raster and vector datasets. Using ArcGIS Explorer, try downloading a vector dataset from one the sources listed previously in this chapter and explore how to change the symbols used on the map and think about how these datasets might work for disaster management tasks. I recommend a technology like ArcGIS Explorer over other free technologies like Google Earth as ArcGIS Explorer has more GIS-specific features such as basic spatial analysis through querying, a wider variety of base map choices so you can explore different cartographic conventions, and integration of standard GIS datasets such as shapefiles and KML than general map-based data viewing like the free version of Google Earth offers.

Working with a tool like ArcGIS Explorer will help you become comfortable with basic GIS operation, for example, adding data layers to a map. Learn how to reorder the map layer drawing levels, explore attribute data associated with vector datasets, and learn general digital map interaction such as panning and zooming. When you are comfortable with these tasks, try working with a more robust GIS technology. In this regard, and depending on your circumstances, options might include trying the 60-day free trial version of ArcMap (http://www.esri.com/software/arcgis/arcgis-for-desktop/free-trial) or QGIS (which was listed previously). If you are student and your initial activities with a technology like ArcGIS Explorer has piqued your interest, look into taking an introductory GIS class (if your university offers one). With a more robust GIS tool, begin to explore the analytical functions of these tools such as buffer, clip, and union (search the help system of the tool you are using for these functions), practice making maps with these tools (see Chapter 2), try creating your own spatial data or editing existing data, and develop metadata for datasets you create. In general, work at becoming more proficient with operating a robust GIS technology; this will allow you to be more competent at using GIS for disaster management tasks (and make you more employable).

For those of you who are looking for ideas on how to set up a *geospatial stack* (i.e., a series of technology combined to provide various functions and services) for disaster management, here are some ideas from the perspective of some of the open-source technologies listed previously. Please note that implementing technology solutions like these requires a substantial amount of IT knowledge in areas such as web servers, databases, system administration, and programming. This is why technologies such as ArcGIS Online and Google cloud-based mapping have become very popular in the past few years as they have eliminated technical barriers to creating a geospatial stack. Thus, if you are unfamiliar with the IT areas needed to create a geospatial stack, make sure to find the right people that can help you if you choose to go this route or consider one of the completely integrated solutions previously mentioned. Also note that the following discussion is not intended to be a definitive guide, but rather a loose set of recommendations for you to follow if you are new to working with open-source geospatial technology in general and want to create your own geospatial stack.

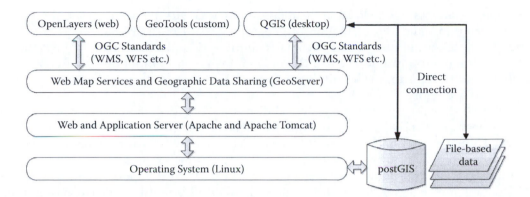

Figure 3.29 An open-source geospatial stack.

A very common and popular combination of technologies to create a stack is the LAMP configuration. *LAMP* refers to the combination of the Linux operating system, the Apache web and application server, MySQL as the database, and PhP as the scripting language. However, each of these components are not set in stone. For example, postgreSQL with postGIS might offer a better choice as a spatial database solution than MySQL, and Python can offer a better choice for scripting related to GIS technology than PhP. Thus, the following list and accompanying Figure 3.29 outlines ideas for an open-source GIS technology stack:

- *Operating system:* Linux (many choices, see: http://www.linux.org/ and http://www.ubuntu.com/download)
- *Web server and application server:* Apache (http://www.apache.org/) and Apache Tomcat (http://tomcat.apache.org/)
- *Web map services and geographic data sharing:* GeoServer (http://geoserver.org/display/GEOS/Welcome)
- *Database:* postgreSQL and postGIS (http://postgis.net/)
- *Web client and presentation:* Open Layers (http://openlayers.org/)
- *Desktop support and data management:* QGIS
- *Custom application development libraries:* GeoTools (open-source Java libraries http://www.geotools.org/), Android API

CHAPTER SUMMARY

In this chapter, you were introduced to GIS on conceptual and technical levels. The chapter began with a discussion about the individual components of the system that comprises GIS. Next, you were shown the various functions that GIS can do such as data and spatial asset management, which is the core of any GIS; analysis to help answer questions and derive insight into spatial problems; programming for developing custom applications and tools to extend the capabilities of GIS; modeling for creating scaled representations of reality and to answer what-if questions; cartography; visualization and map production, which connects modern-day GIS with the millennia-old practice of map making and representation of geographic features; and geocoding, which is the idea of taking textual inputs such as an address or a place name and converting it to coordinates. You also learned a little bit about what GIS cannot do—points that are important to keep in mind as you learn more about GIS and the need to manage technology expectations.

The chapter then provided a technical discussion of GIS data models, which are the ways geographic and spatial entities are represented in digital formats. Specifically, you learned about the two most common data models—vector data, which represents discrete entities as points, lines, and polygons, and raster data, which represents continuous entities as a grid of pixels with varying spatial resolution. Of great importance to all types of GIS data is metadata, or data that describes the contents, structure, lineage, or anything else pertinent to a dataset. Metadata is vitally important for determining the usefulness and relevance of a dataset for a given task or application.

The chapter then discussed specific technologies relevant to disaster management applications. As stated several times in this chapter, this is one part of the book where you will need to be careful to check for updates on specific URLs listed because technology is always changing. However, the specific technologies listed such as Esri, Google and QGIS were chosen given their stability, popularity, and wide user communities. A list of free and open-source datasets relevant to disaster management were also provided to give you some ideas for where to find both reference and thematic data that could be of use for disaster management tasks.

Finally, the chapter gave you some ideas on what to think about when deciding which technologies to use for disaster management and how to get started working with GIS technology if you are new to GIS and learning GIS independently (i.e., you are not taking classes at a university of some other type of training). The chapter concluded with a loose set of guidelines for building a GIS technology stack using open-source GIS technology, if you are interested in the more technical aspects of building a complete GIS solution to support disaster management tasks.

In the next chapter, the relationship between disaster management and GIS is discussed in more detail starting with an overview of the general disaster management cycle (response, recovery, mitigation, planning), the role of GIS within disaster management policy at different scales within the United States (i.e., town, county, state, and federal), and how the international community such as the United Nations and other entities engage in international disaster risk reduction, response, and recovery.

DISCUSSION QUESTIONS

1. What, if anything, might you add to the components of GIS in terms of disaster management areas?
2. With the further increase of 3D virtualization worlds like Google Earth, do you think the concept of map layers is still relevant for disaster management?
3. What other types of GIS data storage formats can you find, besides the ones listed in this chapter, that are used for disaster management applications?

4. From what you've learned about GIS and disaster management so far, what other GIS disaster management analysis scenarios you can think of?

5. What other GIS limitations you can think of and how might those limitations be relevant to disaster management?

6. What are some disaster management features you would represent as vector? As raster? Are there disaster management cases you can think of where either raster or vector could be used?

7. Why is metadata important?

8. If you could use any technology, which GIS technologies might you use for specific disaster management tasks and why?

9. Try downloading some of the datasets listed at the end of this chapter. What were your experiences? For example, were the datasets easy to find and download, or did you have problems?

RESOURCES NOTES

During the creation of this book, the great Dr. Roger Tomlinson passed away. See http://www.npr.org/2014/02/13/276522411/tech-innovator-and-master-of-maps-dies-at-80 for a discussion of this important person in the history of GIS.

For more information about the Dual Independent Map Encoding (DIME) format, see http://www.census.gov/history/www/innovations/technology/dual_independent_map_encoding.html.

For information about the Esri File Geodatabase format, see http://www.esri.com/news/arcuser/0309/files/9reasons.pdf.

More information on ERDAS IMAGINE, see http://www.hexagongeospatial.com/products/ERDAS-IMAGINE/Details.aspx.

For more information on ENVI, see http://www.exelisvis.com/ProductsServices/ENVI/ENVI.aspx.

For examples of Imagery Services available from the US National Map, see http://viewer.nationalmap.gov/example/services/serviceList.html.

For more information on application development with Esri technology, see https://developers.arcgis.com/en/.

For more information about Tiled Map Services, see http://wiki.osgeo.org/wiki/Tile_Map_Service_Specification.

For more information on GeoJSON, see http://geojson.org/.

For more information on the Shuttle Radar Topography Mission (SRTM), see http://www2.jpl.nasa.gov/srtm/.

For more information on Global 30 Arc-Second Elevation (GTOPO30), see https://lta.cr.usgs.gov/GTOPO30.

For more information on Global Digital Elevation Model (ETOPO2), see http://www.ngdc.noaa.gov/mgg/fliers/01mgg04.html.

For more information on Android location services, see http://developer.android.com/reference/android/location/package-summary.html.

For more on HTML5, see http://www.w3schools.com/html/html5_intro.asp.

REFERENCES

Batty, Michael, Andrew Hudson-Smith, Richard Milton, and Andrew Crooks. 2010. "Map mashups," Web 2.0 and the GIS revolution. *Annals of GIS* 16 (1):1–13.

Chrisman, Nicholas. n.d. *History of the Harvard Laboratory for Computer Graphics: A Poster Exhibit*, http://isites.harvard.edu/fs/docs/icb.topic39008.files/History_LCG.pdf (accessed May 25, 2014).

Environmental Systems Research Institute. 1998. Shapefile technical description. http://www.esri.com/library/whitepapers/pdfs/shapefile.pdf

Environmental Systems Research Institute. n.d. "Success stories," Esri, http://www.esri.com/industries/public-safety/emergency-disaster-management/success-stories (accessed December 31, 2013).

Field, Kenneth, and James O'Brien. 2010. "Cartoblography: Experiments in using and organising the spatial context of micro blogging." *Transactions in GIS* 14:5–23.

Kataoka, Mike. 2007. *GIS for Homeland Security*. Redlands, CA: Esri Press.

Liu, Shopia B., and Leysia Palen. 2010. "The new cartographers: Crisis map mashups and the emergence of neogeographic practice." *Cartography and Geographic Information Science* 37 (1):69–90.

Maguire, David J., Michael Batty, and Michael F. Goodchild. 2005. *GIS, Spatial Analysis and Modeling*. Redlands, CA: ESRI Press.

Open Geospatial Consortium (OGC). 2014. "KML," OGC, http://www.opengeospatial.org/standards/kml (accessed April 2, 2014).

Parry, Wynne. 2013. "Why disasters like sandy hit the elderly hard," Live Science, March 8, http://www.livescience.com/27752-natural-disasters-hit-elderly-hard.html (accessed January 17, 2014).

Saul, Michael Howard. 2012. "Few check into city shelters," *Wall Street Journal*, October 29, http://online.wsj.com/news/articles/SB10001424052970204840504578086660466995862 (accessed January 17, 2014).

Schradin, Ryan. 2013. "The benefits of geocoding in the federal government: An exclusive interview with Pitney Bowes Software's Brian Perrotta," EngageGovToday, February 25, http://engage-today.com/gov/the-benefits-of-geocoding-in-the-federal-government-an-exclusive-interview-with-pitney-bowes-softwares-brian-perrotta/ (accessed April 2, 2014).

Standard Unified Modeling Mapping and Integration Toolkit. n.d. "What is SUMMIT?" SUMMIT, https://dhs-summit.us/ (accessed April 2, 2014).

Tomaszewski, Brian. 2003. "Emergency response and planning application performs plume modeling." *ArcUser*, 10–12.

Tomlinson, Roger F. 2007. *Thinking About GIS: Geographic Information System Planning for Managers*. Redlands, CA: ESRI Press.

Valacich, Joe, and Christoph Schneider. 2010. *Information Systems Today: Managing in the Digital World*, Prentice Hall, Upper Saddle River, New Jersey.

van Aardt, Jan, Donald McKeown, Jason Faulring, Nina Raqueño, May Casterline, Chris Renschler, Ronald Eguchi, David Messinger, Robert Krzaczek, and Steve Cavillia. 2011. "Geospatial disaster response during the Haiti earthquake: A case study spanning airborne deployment, data collection, transfer, processing, and dissemination." *Photogrammetric Engineering and Remote Sensing* 77 (9):943–952.

4

Disaster Management and Geographic Information Systems

CHAPTER OBJECTIVES

Upon chapter completion, readers should be able to

1. discern the difference between different disaster management terms,
2. understand different disaster management cycle components,
3. understand the role of Geographic Information Systems (GIS) within different disaster management policies and jurisdictional levels, and
4. understand how international disaster management operates and the various organizations and mechanisms involved in international disaster management.

INTRODUCTION

Up to this point in the book, the discussion has focused primarily on GIS topics, such as the principals of geographic information and maps (Chapter 2) that underlie specific GIS concepts and specific technology (Chapter 3). This chapter closely explores the relationship between disaster management and GIS to give you a deeper understanding of the disaster management application domain. Ideally, by better understanding the characteristics and nature of disaster management, you will get a better sense of how the various system parts of GIS (and not just GIS technology itself) fit within disaster management practice. This chapter assumes you have little to no background in disaster management. Furthermore, this chapter is not a comprehensive discussion of all aspects of disaster management. Thus, the disaster management discussions presented in this chapter are purposely steered toward the role and relationship of GIS with disaster management.

The chapter begins with an overview of the concept of the disaster management cycle—a well-established paradigm that disaster management practitioners and researchers use to understand different disaster phases. The relationship between specific disaster phases and GIS is addressed at length in separate chapters that follow this chapter.

The next chapter part discusses the role of GIS within disaster management policy and practice. The Incident Command System (ICS) and how GIS fits into the ICS is discussed to give you a sense of where GIS fits within broader disaster management policy in the United States. The chapter then discusses how GIS fits with disaster management practice at different governmental scales, starting with local governments (i.e., towns and counties), state-level government, and then federal-level government. Again, although written from a United States perspective, the discussion should give you a sense of how GIS functions across different jurisdictions. The role of the private sector with GIS and disaster management practice is then discussed as the private sector is often very intertwined with the government at all scales through activities such as GIS consulting. The last part of the chapter looks at the international disaster management community and GIS. As mentioned in Chapter 1, as disasters continue to escalate in scale and intensity, more and more, disaster impacts are being felt worldwide demanding a greater need for involvement from the international community. First discussed in this section are international nongovernmental organizations (NGOs) and groups specifically involved with GIS and disasters. Next discussed are international disaster management support mechanisms and organizations specifically involved with disasters and geographic information. This chapter section concludes with a discussion of various United Nations organizations involved in disaster management and GIS. Wherever possible, extensive interviews with disaster management practitioners who work with GIS are provided throughout the chapter to give you a sense of the real GIS for disaster management work people are doing. The following section presents the concept of the disaster management cycle.

DISASTER MANAGEMENT CYCLE

Terms: Emergency, Disaster, Crisis, and Catastrophe

Much like the Chapter 2 discussion on the differences between the terms *data* and *information*, the terms *emergency, disaster, crisis,* and *catastrophe* are often used interchangeably, but they are not the same things. This is a particularly important point in the context of GIS as it is the *geographic scale* that often distinguishes the difference.

An *emergency* is small in geographic scale and can be handled by local officials such as police and fire. For example, a house fire, car accident, or power outage (Lighthouse Readiness Group, 2012). Mapping needs are generally limited to just showing the immediate vicinity of where the emergency is for situation awareness. A *disaster* is larger in geographic scale. The key distinction between an emergency and a disaster is that the disruption caused by the disaster is greater than the local capacity to cope with the event, thus involving resources and officials at multiple levels such as local and state officials (United Nations Office for Disaster Risk Reduction [UNISDR], 2007). A disaster example would be Hurricane Sandy of 2012 as this event clearly overwhelmed the coping capacities of affected communities (Federal Emergency Management Agency [FEMA], 2013). Disaster mapping is much more complex and diverse given the scale and scope of societal disruption that a disaster causes. A *crisis* is often thought of in a temporal aspect, or more specifically, events that lead to a dangerous situation. A crisis example might be a group of elderly people who cannot leave their homes for an evacuation shelter when a hurricane is approaching. A *catastrophe* is like a disaster but bigger in terms of the impacts to physical, social, and organizational systems (Quarantelli, 2006). Hurricane Katrina was considered a *catastrophe* due to the sheer impact on the built environment and need for the federal government to take a significant role in managing the disaster due to inabilities of local and state officials (Phillips, Neal, and Webb, 2012). Catastrophes such as the 2013 Super Typhoon Haiyan in the Philippines also demonstrated how international mapping and GIS support is needed when local GIS capacities are nonexistent or overwhelmed (Lighthouse Readiness Group, 2012; MapAction, 2013).

Disaster Management Cycle

From a disciplinary perspective, geography has traditionally focused on aspects of hazards research, or the *potential* for catastrophic events to occur, and has included topics such as risk (Kunreuther, 2002), vulnerability (Cutter and Emrich, 2006) and mitigation (Mileti, 1999). Research on disasters, or the *realization* of a catastrophic event has tended to be in the domain of psychology, sociology, and more recently, information technology (Information Systems for Crisis Response and Management [ISCRAM], 2014).

As we have seen so far in this book, disaster research within geography has focused on the role of GIS and related technologies in the response phase (Kevany, 2003). However, GIS is relevant to all aspects of disaster management. Disaster management is interpreted here as resource organization and management of activities related to the disaster cycle, or preparing for, responding to, recovery from, and mitigating against disasters (International Federation of Red Cross and Red Crescent Societies, n.d.; FEMA, 2013). Although variations exist on the specific disaster cycle category descriptions, these four are the most common used and relevant to discussion in this book. Figure 4.1 visually demonstrates the idea of the disaster cycle and how GIS underlies each disaster cycle phase.

GIS can play active roles in and across each disaster cycle phase. For example, starting with *preparedness*, or actions taken prior to a disaster with the intent of ensuring a better event response, GIS can be incorporated into technological training such as showing a first responder how to use mobile GIS technologies such as Global Positioning System (GPS)-based smartphone mapping applications or providing citizens with maps of emergency shelters and evacuation routes. Preparedness can also include planning activities such as building GIS hardware, software, datasets, and training capacity for when an event does occur, so that GIS is ready for operational and decision support. In *response*, or actions taken immediately before, during, and after an event to alleviate suffering and prepare for recovery, GIS is critical to supporting situation awareness such as geographic information dissemination like satellite imagery like shown for the 2010 Haiti earthquake in Chapter 3 (Nourbakhsh et al., 2006). For *recovery* efforts, or the rebuilding or improvement of disaster-affected areas, GIS can be incorporated through the use of maps as the objects of collaboration in community planning dialogues and rebuilding efforts (MacEachren, 2005). During *mitigation* activities, or the improvement of the built and social environment in order to reduce, withstand, or prevent disaster impacts, GIS can be incorporated into roles such as mapping, visualization, and identification of vulnerable and at risk populations such as

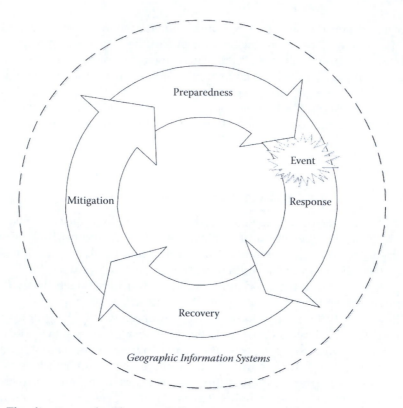

Figure 4.1 The disaster cycle. The actual disaster event occurs between the preparedness and response stages. GIS is relevant and essential to each phase as shown by the outer circle surrounding each phase.

elderly people. Often, the term *risk management* is used to describe what can be considered mitigation activities (UNISDR, n.d.).

Regardless of how GIS is utilized, one theme that emerges from any use of GIS within any disaster cycle phase that GIS serves an *information management* role within disaster management activities. Information management can be loosely defined as the management and collection of information from multiple sources and the dissemination of that information to multiple audiences (Griffiths, 2006).

ROLE OF GIS WITHIN DISASTER MANAGEMENT POLICY AND PRACTICE

The following sections discuss the role of GIS within disaster management policy and practice to give you a better sense of how specifically GIS functions within government, private sector, international, and other organizations at multiple geographical and jurisdictional scales.

Policy in the United States: The National Incident Management System (NIMS)

The National Incident Management System (NIMS) is an incident management template scalable to all incident types that is designed to guide government, nongovernment, and private sector organizations on all aspects of disaster management (response, recovery, planning, and mitigation) for reduction of loss of life and property and environmental damage. (United States Department of Homeland Security, 2008). NIMS is not disaster response or a communication plan, targeted to one type of incident and specific incident response personnel (United States Department of Homeland Security, 2008). Subsequent book chapters will discuss other FEMA policy frameworks specifically related to response, recovery, and mitigation. To integrate incident response and emergency management practice, five key areas are the focus of NIMS (United States Department of Homeland Security, 2008):

I. Preparedness
II. Communications and Information Management
III. Resource Management
IV. Command and Management
V. Ongoing Management and Maintenance

Of particular interest to GIS and these five areas are components I (Preparedness) and II (Communications and Information Management). Component I (Preparedness) outlines "specific measures and capabilities that emergency management/response personnel and their affiliated organizations should develop and incorporate into their overall preparedness programs to enhance the operational preparedness necessary for all-hazards emergency management and incident response activities" (United States Department of Homeland Security, 2008, 9). Five subareas are defined within the preparedness component to achieve preparedness. One of those five is *mitigation*, which outlines guidelines for risk reduction through activities (many of which are spatial in nature), such as public education and outreach, building code enforcement, evacuation zone planning, and a direct reference to GIS (text in italics added by the author of this book): "periodic remapping of hazard or potential hazard zones, *using geospatial techniques*" (United States Department of Homeland Security, 2008).

Component II (Communications and Information Management) emphasizes the importance of "flexible communications and information systems that provide a common operating picture to emergency management/response personnel" (United States Department of Homeland Security, 2008, 23). *Common operating picture* (COP) is a concept similar to situation awareness (SA), which you learned about in Chapter 1, but with the slight difference that it is the situation awareness for *all parties* involved in an incident and not just a single person or party (FEMA, 2009). Much like SA, COP is constantly updated as a situation changes and is fed from multiple information sources such as traffic, weather, voice, available resources, and more (United States Department of Homeland Security, 2008). Geographic maps are often the basis for grounding all of the elements that comprise a COP. Component II outlines how incident information is to be used by organizations and operations practice to inform decision making. The incident information section of Component II specifically describes *geospatial information* as incident information that can be generated for decision-making purposes. The following is the text about geospatial information from the NIMS; concepts and ideas you have learned about in this book are *italicized* to give you a sense of how those concepts are directly related to official IS disaster management policy(United States Department of Homeland Security, 2008, 28):

> *Geospatial information is defined as information pertaining to the geographic location and characteristics of natural or constructed features and boundaries.* It is often used to integrate assessments, situation reports, and incident notification into a common operating picture and as a data fusion and analysis tool to synthesize many kinds and sources of data and imagery. The use of geospatial data (and the recognition of its intelligence capabilities) is increasingly important during incidents. Geospatial information capabilities (such as *nationally consistent grid systems or global positioning systems based on lines of longitude and latitude*) should be managed through preparedness efforts and integrated within the command, coordination, and support elements of an incident, including resource management and public information.
>
> *The use of geospatial data should be tied to consistent standards,* as it has the potential to be misinterpreted, transposed incorrectly, or otherwise misapplied, causing inconspicuous yet serious errors. Standards covering geospatial information should also enable systems to be used in remote field locations or devastated areas where telecommunications may not be capable of handling large images or may be limited in terms of computing hardware.

Incident Command System (ICS)

The ICS is outlined under component IV (Command and Management) of NIMS. Although GIS is not explicitly mentioned in NIMS for ICS, ICS is nonetheless an important emergency management concept that is widely used (in the United States) for incidents at all organizational and jurisdictional levels. Grounded in the NIMS principle that although most incidents start and end at the local level, when incidents begin to expand in terms of geographical area, resource needs, disciplinary needs, or jurisdictions, the ICS "provides a flexible core mechanism for coordinated and collaborative incident management, whether for incidents where additional resources are required or are provided from different organizations within a single jurisdiction or outside the jurisdiction, or for complex incidents with national implications (such as an emerging infectious disease or a bioterrorism attack)" (United States Department of Homeland Securitym 2008, 45). Figure 4.2 outlines the general, high-level structure of the ICS command and general staff.

Discussion of all the components shown in Figure 4.2 are beyond the scope of this book (see United States Department of Homeland Security, 2008. for details). However, of particular note in terms of the specific connections between the ICS, command and general staff, and GIS is the planning section, which is "responsible for collecting, evaluating, and disseminating operational information pertaining to the incident ... [and] ... prepares and documents Incident Action Plans and *incident maps*, and gathers and disseminates information and intelligence critical to the incident" (United States Department of Homeland Security, 2008, 103), (emphasis added by author of this book). Maps in particular are the responsibility of the situation unit within the planning section. Note how the language of Component II (Communications and Information Management) discussed previously matches closely with the descriptions of the specific tasks conducted by planning section situation units.

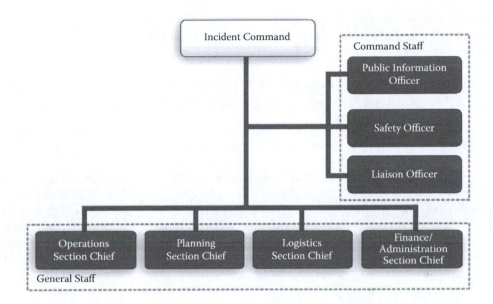

Figure 4.2 The Incident Command System Command and General Staff. (From United States Department of Homeland Security. "National Incident Management System," United States Department of Homeland Security, http://www.fema.gov/national-incident-management-system [accessed May 25, 2014].)

The ICS can also incorporate a wide variety of *technical specialists,* or people who serve specific technical functions depending on the incident, into various parts of the ICS structure. For example, a legal specialist can be assigned to the finance section to cover financial matters, or a legal specialist can be assigned to the command staff to provide legal advice on matters such as mandatory evacuation orders or media access restrictions (United States Department of Homeland Security, 2008). The technical nature of GIS has accordingly led to the creation of *Geographic Information System (GIS) Specialist* positions to work within the ICS, and to ideas for you to keep in mind if you are looking for employment in the GIS disaster management domain. For example, the United States Department of Labor Occupational Safety and Health Administration (OSHA) states: "The Geographic Information System (GIS) Specialist is responsible for gathering and compiling updated spill information and providing various map products to the incident. The GIS team will work with the Situation Unit and the Information Management Officer to ensure accurate and rapid dissemination of oil spill information to the Incident Command System (ICS)" (United States Department of Labor, n.d.). A review of the knowledge, skills, and abilities for these positions reveals many topics covered in this book such as reference maps, coordinate systems, map projections, and other technical topics covered later in this book (see the Resources section of this book for links to ICS GIS-related positions).

The following sections discuss GIS disaster management within the context of various levels of government. A running example from New York State and the United States is used to illustrate specific examples at different jurisdictional scales. These examples are simply meant to illustrate specific cases, and you are encouraged to review your own governments wherever you live to find comparable examples of GIS at the same jurisdictional scales.

United States Department of Homeland Security (DHS) Geospatial Concept of Operations (GeoCONOPS)

The United States Department of Homeland Security Geospatial Concept of Operations (GeoCONOPS) is designed to coordinate geospatial response activities and geospatial communities involved in federal-level emergency management under Presidential Policy Directive 8 (PPD-8), which includes individual Emergency Support Functions (ESFs), the Joint Field Offices, FEMA Regional Coordination Centers (RRCCs), and the National Response Coordination Center (NRCC) (United States Department of Homeland Security, 2013, 4). GeoCONOPS can be considered part of broader calls for a "national framework for geospatial information sharing that links policy to collaborative governance that is aligned to mission and business functions with an emphasis toward common geospatial data, shared capabilities and infrastructure, and an interoperable architecture that supports standards and innovation" (Alexander, 2013). GeoCONOPS "ensures that timely and accurate geospatial data is shared across the entire geospatial community resulting in better informed decision making across all phases of an incident (United States Department of Homeland Security, 2013, 5).

The GeoCONOPS Community Model (Figure 4.3) is a graphical representation of the GeoCONOPS framework. The model accomplishes the following (United States Department of Homeland Security, 2013, 10):

- Identifies actors and stakeholders that support the geospatial community mission
- Identifies the information environment and actor responsibilities
- Documents information sharing within and outside the geospatial community
- Illustrates high-level processes across the geospatial mission operations and the correlating relationships of these processes with stakeholders

United States National Spatial Data Infrastructure

Closely related on a conceptual level to GeoCONOPS, the United States National Spatial Data Infrastructure (NSDI) has the goal to "to reduce duplication of effort among agencies, improve quality and reduce costs related to geographic information, to make geographic data more accessible to the public, to increase the benefits of using available data, and to establish key partnerships with states, counties, cities, tribal nations, academia and the private sector to increase data availability" (Federal Geographic Data Committee, 2007). Spatial data infrastructures (SDIs) are not unique to the US federal government. Calls for the development of SDIs with large governments and organizations to promote geospatial dataset sharing, access, and interoperability have been made for many years (see Masser, 2005; Bernard et al., 2005). Examples of other major SDIs include:

- Infrastructure for Spatial Information in the European Community (INSPIRE): http://inspire.ec.europa.eu/
- United Nations Spatial Data Infrastructure (UNSDI): http://www.ungiwg.org/content/united-nations-spatial-data-infrastructure-unsdi

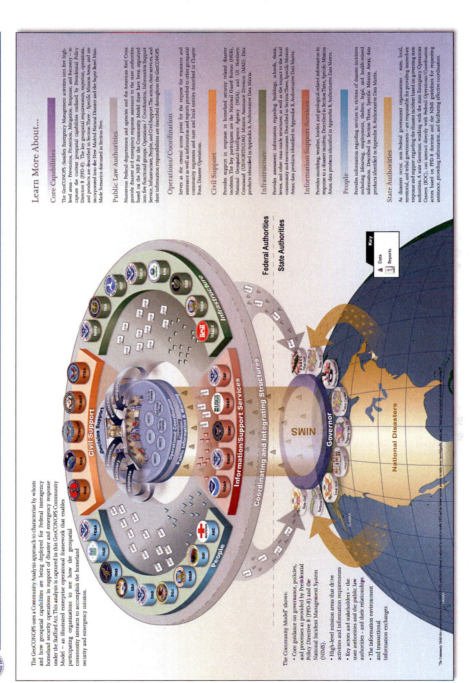

Figure 4.3 The GeoCONOPS Community Model. Department of Homeland Security, Homeland Security Geospacial Concept of Operations (GeoCONOPS): Coordinating geospatial support for the Homeland Security Mission, Version 5.0, June 2013.

Local Government: Cities, Towns, and Counties

Within the context of local governments in the United States, such as cities, towns, and counties, GIS plays an important role across multiple agencies. Often, the level of GIS activity within local governments is tied into the size of the jurisdiction and the amount of funding available and needed for GIS. For example, smaller towns may not have a need for GIS capabilities. They might rely on any GIS and mapping needs using external consultants or data and map services provided by their respective counties or states. In fact, some small towns may not even be aware of GIS, and instead rely on well-established CAD (computer-aided drafting) services for mapping needs based on existing relationships with engineering consultants that often provide maps for municipal transportation, water, and sanitation projects.

At the county level in the United States, GIS often exists as its own division or as part of the broader county division. Again like smaller towns, counties with smaller populations and a subsequently lower tax base, or counties without a major city may have less GIS capacity and more reliance on external GIS consultants. In terms of disaster management, GIS divisions are often critical for a variety of data and services needed. For example, as you saw in Chapter 1, during the 9/11 attacks, GIS was essential to responding to the terror attacks and was even more challenging due to the fact that the GIS center in one of the World Trade Center towers was destroyed. Often in major cities, GIS divisions need to work very closely where critical infrastructure entities such as utilities and transportation exist. Furthermore, city GIS units work very closely with state and federal agencies to coordinate disaster management activities. Hurricane Sandy of 2012 was an excellent example of this when the subway tunnels of New York were flooded with storm surge requiring close coordination with GIS data from the city, the metropolitan transportation administration (MTA), and FEMA (Lewis, 2013; ArcGIS, 2014).

County GIS: Interview with Scott McCarty

Scott McCarty (Figure 4.4) is the GIS Operations Manager for the Monroe County, New York, GIS Services Division* (which also maintains the GIS truck shown in Chapter 1). He attended college at SUNY Brockport and received a bachelor's degree in biology. His career with the county began in 1992 with the Department of Environmental Services, testing wastewater and industrial waste. In 1998, he shifted duties and began using GIS to digitize sewer record maps. This was at the same time when other GIS initiatives started within the county, such as the real property tax map parcel conversion. Also at the same time, the Planning and Health Departments were ramping up their mapping efforts. As interest across the county grew, in 2000 a major consolidation was developed when three employees from the DES Pure Waters side (including Scott) and two employees from the Planning Department were formed into the GIS Services Division. To this day, the GIS Services Division continues to work with other county departments and local towns and villages to support their GIS needs. The following is the first of a two-part interview conducted for this book with Mr. McCarty in February 2014. In this portion of the interview, he answers questions about his specific GIS work with Monroe County related to disaster management. The second half of this interview is presented in Chapter 9 where Mr. McCarty provides advice on getting a job in the GIS industry for disaster management and the future of GIS for disaster management.

---

* Monroe County website, http://www.monroecounty.gov/gis.

Figure 4.4   Scott McCarty.

What types of GIS disaster management activities does Monroe County GIS do?
We participate in several different public safety exercises throughout the year. We have a major involvement with the emergency operations center [EOC] in Monroe County. When the EOC is activated, we are called in to provide mapping support to various agencies, either through hard-copy maps or web-based mapping applications. We have heavy involvement with the EOC during live events, as well as planned exercises. One exercise that we are involved with on an annual basis is a border patrol exercise that encompasses the shoreline along Lake Ontario. Agencies involved in this exercise include the US Border Patrol, US Coast Guard, and various other state and federal organizations.

We also participate in two exercises each year that involves the local nuclear power plant. There are different scenarios given to the participating agencies each year to help prepare our region in the event that a real situation occurs. These scenarios could be in the form of a hostile takeover, or it could be weather related. Weather-related exercises such as flooding, hurricane, and ice storms are also on the annual EOC agenda. Whatever the case may be, we react to the information that is given to us.

So, is the terminology from the Incident Command System a big part of these exercises?
Yes. We follow the lead of emergency management and the public safety agencies. They know what we can provide and they know what data we have. We have certain things set up ahead of time, obviously, because you really never know when a live event is going to happen. We want to be prepared, so it is important that we generate base maps and data layers prior to these exercises and live events. Having these base maps and specific datasets related to flooding, hurricane, or ice storms already created, it allows us to easily incorporate them into the EOC operations. The same goes with the nuclear power plant. There are certain evacuation zones that are already set up, as well as siren locations and evacuation routes. They're all determined ahead of time.

Tell me about your specific GIS disaster management activities at the county.

An important item leading up to these events or exercises is preplanning. Preplanning is very important to Monroe County's Office of Emergency Management. And so, specifically what we do in the GIS division is we make sure that we can provide the information when it's asked for. That, I guess, is one of the biggest fears is that you can't provide something when asked. So, we spend a lot of time developing datasets and applications ahead of time. At the EOC we provide mapping and GPS support to first responders and EOC staff. And when I say GPS support, that's where we get into the mobile command unit (discussed in Chapter 1). With GIS support, we're starting to move towards web-based mapping. Instead of providing hard copy maps, we now like to push the data out to web-based applications, which is something that doesn't require everybody at the EOC to have an ArcMap license. Web-based mapping apps are going to play a major role with what we do with anything involved as far as these types of events.

In terms of my day-to-day activities, I am *still a hands-on technical person*. I'm still doing projects and at the same time managing the division. In terms of tools, yes, we maintain Esri on the desktop and server side. We just recently purchased a web mapping platform called *GeoCortex** that we're implementing as we speak. All of our existing web apps right now are built in *Flex,*† which were developed by a consultant. This was always something that we wanted to do in-house, but we've just never been able to either hire a developer or have the time to go take the training ourselves. We've been using this software called GeoCortex, which makes it real easy for the end user to develop web maps.

How does Monroe County GIS interact with private sector, local governments, and state and federal GIS entities in terms of disaster management activities?

We do a lot of preplanning with these groups. Let's take the border patrol exercise as an example. We'll get together with all the towns that are on the edge of Lake Ontario prior to an event, and we'll come up with different maps or datasets that might be useful during an exercise or a live event. For example, we keep in close contact with the local police departments, such as the town of Greece (located in Monroe County), who might be involved in these types of activities. Another example would be local utility companies, especially when there's an activation and are going to be at the Emergency Operations Center. Local municipalities and school districts are others that we work closely with. We certainly reach out to them for these types of preplanning events. We're also involved with them with regards to training. Monroe County provides GIS and GPS training to all of its municipalities. Actually, any resident in Monroe County can take our Intro to GIS training class at no cost.

We've put a lot of effort into training local municipalities and other departments mainly because in the past when we first started as a division, we would end up doing a majority, if not all of the work. As an example, a local department of public works [DPW] would come to us asking to map their light pole locations. Our response would be, "Okay—we'll do it for you or we'll come out and you

---

* GeoCortex website, http://www.geocortex.com/.

† Adobe, http://www.adobe.com/products/flex.html.

can send one of your guys around with our guy and make a map of light poles."
As time went on, we just weren't able to keep up with the workload because of
the popularity of getting data into a GIS. So, we put a lot of effort into a training
class geared towards someone who hasn't had any GIS experience. And so, at the
end of this class they could at least, maybe with our help, get a project started,
know how to build a map, query data, maybe bring GPS data into their projects,
and kind of send them on their way, but, of course, always be here for technical
support. They can certainly call us at any time and we'll give them a hand.

We do our best to provide training and get people started on projects. We'll provide the
municipality with a GPS receiver if they have the personnel to do the initial
data collection. Once that is complete, we'll help them take the data off of the
receiver and get it into a GIS format. We have a good relationship with municipalities. Much like counties vary in how far ahead some are than others as
far as GIS, it's the same with the towns and villages within Monroe County.
There's some towns that have designated GIS people and there's some towns
that don't have anybody. Then, there's a lot of people in the middle.

Most major cities in the United States have dedicated GIS divisions. For example, NYC
[New York City] has a GIS unit within its information technology and communications
division. The GIS division is responsible for maintaining a city base map and a variety of
tools and datasets support other aspects of city government for public safety, analysis, and
policymaking (NYC Information Technology and Communications, 2014).

State

In the United States, all states maintain GIS capabilities to varying degrees. Continuing
with the example of New York State, statewide GIS services are located within the Office
of Information Technology Services (ITS). Of particular note to New York State GIS and in
other state-level GIS offices in the United States, is the New York State GIS Clearinghouse
(https://gis.ny.gov/). GIS clearinghouses in general were first mentioned in Chapter 3.
They are very important sources for downloading state-specific GIS data layers. Of
particular note in this regard for New York State is the dissemination of digital orthoimagery annually collected by New York State and provided for free through the GIS
clearinghouse. Additionally, the clearinghouse provides access to state-specific GIS data
layers created by the federal government such as United States Geological Survey (USGS)
digital raster maps (that you first saw in Chapter 1) for New York State. Furthermore,
they provide a forum for a statewide community of GIS users to share specific data from
their town, village, county, or other organizational activities in either an open-access,
available-to-download format or through a secure, password-protected, members-only
access format for datasets that have restrictions. Finally, statewide GIS clearinghouses
are an important location for the GIS community to interact with one another; for example, providing contact information for clearinghouse members or information on events
such as GIS conferences and training sessions. In terms of disaster management, state-
level GIS offices will also coordinate closely with critical infrastructure entities and the
federal government to provide GIS datasets and analysis to support a variety of disaster
management activities.

National

FEMA

In the United States, at the national level, FEMA is perhaps the best example of a government entity that integrates GIS for disaster management. Maps, mapping, and GIS are critical to many of FEMA's activities as you learned previously in this chapter in the NIMS discussion. Discussion of all aspects of GIS and FEMA is a vast topic that would require an entire book unto itself. As a starting point to learn about FEMA in GIS, review the FEMA Enterprise GIS Services web page (http://gis.fema.gov/) where you can find specific information about GIS and its support role in planning, preparing, recovering, and rebuilding activities, and GIS data feeds such as currently declared and historical disasters and emergencies. Furthermore, take a look at online web-based mapping tools that FEMA provides such as the FEMA GeoPlatform (Figure 4.5) to get a sense of the breadth of FEMA GIS activities.

GIS and Other US Federal Agencies

Although FEMA is the primary federal agency that uses GIS for disaster management activities, many other federal agencies use GIS for disaster management, and in particular, for international incidents. Of particular note in this regard are the Humanitarian Information Unit (HIU; Figure 4.6) of the US State Department, the US Agency for International Development (USAID) GeoCenter, and the National Geospatial-Intelligence

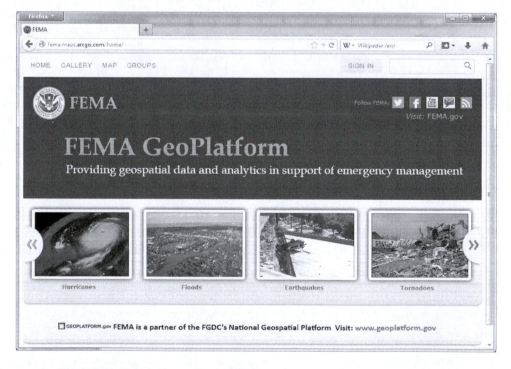

Figure 4.5 The FEMA GeoPlatform. (From FEMA website, http://fema.maps.arcgis.com/home/.)

Figure 4.6 The Humanitarian Information Unit. (From HIU website, https://hiu.state.gov/Pages/Home.aspx.)

Agency (NGA) geoanalytics activities (Locker, 2013). The HIU "is an inter-agency office providing U.S. Government officials and other aid organizations with geographic data and analysis to prepare for and respond to complex humanitarian emergencies worldwide" (US Department of State, n.d.).

Particular activities of the HIU as of 2013 include crowdsourcing of high-resolution imagery to determine refugee locations in Syria and development of international boundary datasets.

The United States Agency for International Development (USAID) is the agency charged with promoting international development through the United States government and similar organizations exist in other developed countries. A good example of this is the Department for International Development (DFID) in the United Kingdom. Although not explicitly tied into disaster management, international development is an important activity for disaster risk reduction. For example, building houses to be more earthquake resilient or developing flood mitigation strategies such as wetland restoration. In 2011, USAID launched the USAID GeoCenter, with the purpose to "enhance USAID's capacity in strategic planning and programming, evaluation, and research with the use of the powerful tools of geospatial analysis" (USAID, 2011). A review of the USAID GeoCenter's ArcGIS.com web page shows the variety of development projects the GeoCenter is involved with a particular emphasis on aid tracking (see link in the Resources section).

Although primarily tasked with supporting military applications of what is known as *geospatial intelligence* or GEOINT, the NGA continues to play an important role in large-scale and international disaster response activities. In particular, the NGA provides access to high-resolution satellite imagery related to geospatial analysis that provides essential situation awareness during a disaster response (Yasin, 2013).

Non-US Federal-Level Disaster Management: Interview with Dr. Michael Judex
The following interview provides perspectives on GIS for disaster management activities from a non-US, federal-level management agency, the German Federal Office of Civil Protection and Disaster Assistance.

Dr. Michael Judex (Figure 4.7) is a Geoinformation Project coordinator at the Federal Office of Civil Protection and Disaster Assistance (German: Bundesamt für Bevölkerungsschutz und Katastrophenhilfe, or BBK; http://www.bbk.bund.de/EN/Home/home_node.html), which is Germany's top federal organization for civil protection. He holds a PhD in geography from the University of Bonn, Germany, with a specialty in GIS and remote sensing and has a variety of international experience working in West Africa on land-use/land-cover issues and a large European project preparing the European emergency mapping service. At BBK, in addition to a variety of geoinformation-related activities, he is Germany's representative to the European Union Copernicus Emergency Management Services (EMS). Since 2012 this is an operational earth observation service for civil crisis management. The following is the first of a two-part interview conducted for this book with Dr. Judex in January 2014. In this portion of the interview, he answers questions about his specific work with BBK and the role of BBK in broader European disaster management activities. The second half of this interview is presented in Chapter 9 where Dr. Judex provides advice on getting a job in the GIS for disaster management industry and the future of GIS for disaster management.

Figure 4.7   Dr. Michael Judex.

What types of disaster management activities does BBK do, and can you generally describe BBK as an organization?

BBK is a federal office. We are a technical administration under the Ministry of Interior. We're responsible for the civil protection at [a] national level. That means we have activities to support all phases of the crisis management cycle such as prevention, preparedness, and disaster response. We also have some operational units to support crisis management activities. That's mainly the joint situation and information center we have at the national level that's within our headquarters building which is the top situation center in Germany for civil protection. We have also a satellite-based warning system in Germany. That's a system where if there's a nationwide danger, we can trigger the warning system. The warning signal is distributed by a satellite within minutes to all TV and radio stations and other authoritative institutions. We are working at the moment to further develop it towards a modular warning system so we can alert even more end devices such as smartphones, home-based smoke detectors, and other devices. Federal states can use the system to issue alerts for local events. That's the idea behind the modular warning system. But I've to say that we have no operational activities on the ground. That's not our responsibility. Our task is mainly to support other institutions in Germany, foremost the Ministry of Interior. But we also support the federal states and counties as political units under the federal states.

Is the BBK equivalent of like FEMA, the Federal Emergency Management Agency, in the United States?

A little bit, yes, as we develop concepts and guidelines for civil protection and we provide a joint situation center at [the] national level. But there are also differences as far as I know. For example, FEMA has some operational activities on the ground during major disasters such as assessment teams, which we do not have. So, we really have a small federal office. We support the German federal states with concepts for GIS technology and provide map-based situation reports, but we do not prescribe a specific GIS for them.

What is your specific work with GIS and disaster management at BBK?

I am responsible for coordinating access to and use of geoinformation in our federal office where we have several departments. In every department, geodata are used. We have really found the need for one geoinformation coordination position. We are now just on the way to build up and implement a spatial data infrastructure for common use. We have a dedicated geospatial information system that is specifically for the operation center. We are currently updating that information system. Another example is risk analysis for which we need a large amount of data. I support them with the data access and geospatial analysis methods. And there's the domain of critical infrastructures. For example, we have drinking water wells for emergency situations that we are now managing with a web-GIS system. A very important task of our federal office is to develop concepts and to support other institutions in using geoinformation and by that also support the use of geoinformation services and data. To do that, we regularly organize workshops with different actors in the field.

As a national federal office, we are the hub to several international mechanisms in civil protection. I'm national representative to the European emergency mapping service within the large Copernicus program. That includes mainly the coordination of user requirements, and support of the general policies in Germany regarding that program.

Finally, I support the joint situation center during times of crisis by coordinating the access to remote sensing imagery. Remote sensing is an important tool that is used as [an] additional information source to improve situational awareness. That's not only satellite imagery, but also aerial imagery, and additionally, imagery from unmanned aerial vehicles which is becoming more and more important. Also the production of special situational maps is part of my job.

How well at the different jurisdictional levels within Germany would you say geoinformation is being used?

That's an interesting question. The usage is quite different at different levels. I would say it's actually being used at the very local level. For example, the major cities and towns in Germany, they all have spatial data infrastructures. Central data warehouses are also used more and more by entities like the fire department because they are also responsible for the civil protection tasks. At the federal state level, it's much more the political steering or the development of policies and they're quite far away from the real usage of geoinformation and to perform analyses, for getting more information, for getting better and quicker insight into situations. So hopefully we will see a development there. We represent the national level and we have at least five geospatial experts and we are doing very different types of geospatial analyses and develop concepts for the use of GIS for risk and crisis management.

Do you ever find cases of what you might call culture clash? For example, people that have been working in disaster management for 20+ years aren't willing to accept new technology or new ideas even though these things are there, or they're not really seeing the value of Geographic Information Systems?

Yes, absolutely, that is sometimes the case. But it's really dependent on the position and the size of the institution. As already said, the major towns in Germany, for example, they have really quite comprehensive geodatabases. They use geoinformation for resource planning, and they use GIS for situation maps and other activities. So, that's there, but if you're moving more into countryside areas, then the situation is more that GIS is not used or even not understood. They're really hesitating to use new technologies, even if you can just demonstrate the benefits. One reason might be that for the introduction of GIS you need also organizational adaptations and dedicated resources.

How do the activities of BBK fit within broader European Union disaster management activities or even perhaps international activities?

Our joint situation and information center is actually the only central communication access point for several international mechanisms in the area of civil protection. The most important one is the common civil protection mechanism of the European Union. That is a coordination mechanism to request foreign assistance during large disasters not only within the European Union, but also at the international scale. So, for example, during the last typhoon in the Philippines, there have been several requests by the Philippine government and they have been coordinated by the European Union Emergency Response Coordination

Center with this mechanism. We coordinate the request in Germany and liaise with humanitarian actors that are deployed. Another important mechanism is the European Copernicus Emergency Management Service as already mentioned. It is an operational earth observation service for civil crisis management. In June 2013 we had a major flood event in Germany and we activated the service to receive satellite-based mapping of several flood-affected areas. Again, the requests are coordinated by the joint situation center.

Another activity in relation to international disaster management is, for example, our academy for civil protection where about 10,000 people are trained each year. Also EU assessment teams are trained at this academy. So we are taking part in the standardized training system for EU experts. In general, people can learn everything from the basic operational things such as "how do I work in a control room," up to various specific topics such as the handling of chemical incidents or medicine treatment trainings. Furthermore, people from around the world come to the training center. So, even though the center is basically for German civil protection people, we have several seminars often with international participation.

Private Sector

As alluded to previously in this chapter, the private sector plays a very important role in GIS for disaster management. Very often, government agencies rely on external GIS contractors for GIS data development, GIS software application development, map production analysis, and a variety of other tasks that government staffs are unable to do. A very typical scenario is that an engineering company that provides civil engineering services will also have a GIS branch within the company. *Pure* GIS consulting companies also exist with a specific focus only on GIS activities. To get a sense of the kinds of private sector companies that provide GIS services, it is recommended you take a look at the GIS jobs clearinghouse at http://www.gjc.org/gjc-cgi/listjobs.pl.

Although not specifically geared toward disaster management as opposed to GIS in general, the GIS jobs clearinghouse should give you a good sense of the kinds of companies there are, and the types of skills they are looking for in potential employees. The site is also useful for reviewing GIS jobs in government. Chapter 9 further discusses getting a job in the GIS for disaster management field.

Private-Sector Perspective: Interview with Alan Leidner

Alan Leidner (Figure 4.8) is a GIS practitioner, advocate, and thought leader with over 40 years of experience working in various aspects of urban planning, information technology (IT), and GIS in both government and the private sector. During his career he has worked for the New York City Department of City Planning, the Mayor's Office of Operations in charge of exploring new technology, and the Department of Environmental Protection, where he became IT director and led the citywide effort to build New York's first photogrammetric and planimetric base maps. He finished his city career in the Department of Information Technology and Telecommunications (DOITT) as Assistant Commissioner and Director of the Citywide GIS Utility. While at DOITT, he directed the Emergency Mapping and Data Center (EMDC) that supported the response community

Figure 4.8   Alan Leidner.

following the 9/11 terrorist attack on the World Trade Center. In 2004 he retired from city government and started working for Booz Alan Hamilton (BAH; http://www.boozallen.com/) a Washington, DC-based consulting firm.

At BAH, Alan has spent much of the last 10 years working in support of the Homeland Infrastructure Foundation Level Data (HIFLD) program (https://www.hifldwg.org/). The HIFLD program was created in February 2002 as a direct result of the 9/11 attack on the World Trade Center when federal authorities recognized that the nation did not have comprehensive information about its critical infrastructure. HIFLD was established to improve the collection, integration, and sharing of infrastructure-related geospatial information across all levels of government, for the purpose of creating a common data foundation to be used for visualization and analysis. Federal agencies that comprise the HIFLD working group include the Department of Defense (Homeland Defense and Americas' Security Affairs [HD&ASA]), the DHS National Protection and Programs Directorate Office of Infrastructure Protection (NPPD OIP), the NGA Office of Americas, the Department of Interior (USGS and National Geospatial Program [NGP]); and FEMA.

HIFLD directs the acquisition and assembly of hundreds of layers infrastructure-related data obtained from federal and commercial sources. The more than 560 datasets that comprise the Homeland Security Infrastructure Program (HSIP) Gold data compilation were sourced from 57 government and private sector organizations and provide seamless nationwide coverages that are registered to a common national base map, so that features on different layers can be accurately related to each other. HSIP Gold is probably the largest such national geodata compilation to be found in the world. As a member of the HIFLD to the Regions (HTTR) program, Alan has focused his efforts on the Northeast Region of the United States covering New York State, New Jersey, Puerto Rico, and the Virgin Islands. He has worked to integrate HIFLD data with often more detailed state and local datasets, and has supported a wide variety of security analyses and responses to disaster events including Hurricane Sandy.

The following is the first of a two-part interview conducted for this book with Mr. Leidner in January 2014. In this portion of the interview, he answers questions about his specific work in the private sector related to GIS for disaster management. The second part of this interview will be presented in Chapter 9 where Mr. Leidner provides advice on getting a job in the GIS for disaster management field and the future of GIS for disaster management.

Tell me a little bit about Booz Alan Hamilton (BAH) and its GIS work?

Booz Allen Hamilton provides management, technology, and security services primarily to federal defense, intelligence, and civilian agencies. It has almost 25,000 employees and is based in the Greater Washington DC area, with offices throughout the US and the world. Thousands of Booz Allen employees work with geographically tagged information (geointelligence) obtained through the use of a wide variety of remote sensing instruments. Many of these employees are GIS technicians and analysts. Booz Allen was selected as the contractor to support the HIFLD program. The Booz Allen HIFLD support staff is largely comprised of GIS practitioners and currently includes about a dozen analysts and programmers.

What are the kinds of GIS-related work you did for the HIFLD program?

Over the course of my six years with the HIFLD program, I was responsible for building networks of federal, state, and local GIS managers in the Northeast Region, many of whom worked for public safety, emergency management, and homeland security–related agencies and private companies. I distributed the Homeland Security Infrastructure Protection [HSIP Gold and HSIP Freedom] data compilations to these network contacts and also identified local and state data holdings for sharing purposes. I worked in support of a large number of vulnerability assessments for important facilities such as major bridges in the NY Metro area and the regional electric power grid; and on national security events including meetings of the UN General Assembly. In addition, I supported responses to a number of emergency and disaster events ranging from Super Storm Sandy and other major tropical and winter storms, to large-scale industrial accidents such as the fire at a Puerto Rico–based refinery and oil storage depot.

So, would you say then you hired a range of different people, like GIS analysts, people that could really handle spatial data, working with Esri products to do the analysis, making maps, and with GPS doing field collection to support?

A number of different GIS-related skills are being utilized in support of the HIFLD program. Certainly the majority of the HIFLD support staff are Esri users and have advanced training in GIS analytics. My own particular skills are not technical, although I am quite knowledgeable about GIS capabilities and am able to use web-based GIS applications. But I am much more of an organizer and coordinator, developing my skills during the ten years it took to win approval for and build NYC's enterprise GIS system. I greatly appreciate the fact that HIFLD leadership understood that one key to the success of their program was the ability to bring people together. Important infrastructure data is in the hands of many different organizations at all levels of government and the private sector. Any credible infrastructure protection program has to be able to tap these information sources, facilitate sharing, and enable collaboration. HIFLD has been doing this job admirably by holding many dozens of information sharing and networking meetings, distributing data products needed by the entire homeland security and law enforcement community, and providing essential information through their website. So, my strength was a good fit for the program and complemented the more technical skills of my colleagues.

INTERNATIONAL DISASTER MANAGEMENT COMMUNITY AND GIS

Our increasingly interconnected and interdependent world combined with risks and hazards ranging from increased population density, overdependence on computing technology, terrorism, climate change, and many other factors continue to demonstrate how disasters are increasingly an international phenomenon and response, long-term recovery, and risk reduction require attention from the broader global community. In addition to the technical, tactical, and cultural challenges faced at all levels within a single country during a crisis, international disaster management (and response in particular) is faced with unique challenges such as

- language, cultural, and social barriers between foreign responders and native populations;
- unfamiliar operational environments for responders (for example, working in a rural desert area when one is used to working in Washington, DC);
- lack of central command-and-control structures to manage and coordinate very large-scale disasters spanning international boundaries;
- operation in countries with unstable political systems;
- working with unclear operational and political jurisdictions;
- problems with effectiveness and relevancy of response coordination and aid offers from foreign governments; and
- lack of situation awareness across large geographic areas of an affected region.

Disaster management practitioners operating at the international scale are thus faced with the enormous challenge of collecting and disseminating information to alleviate human suffering within these and other constraints.

The following sections discuss some of the important organizations and mechanisms for international disaster management.

Nongovernmental Organizations

MapAction
MapAction (http://www.mapaction.org/) is an international NGO with a specific focus and capacity to deploy internationally to disaster sites to set up a rapid mapping and geographic information acquisition and analysis within the early stages of a disaster response. More specifically, and in their own words, "MapAction delivers this vital [disaster] information in mapped form, from data gathered at the disaster scene. Creating a 'shared operational picture' is crucial for making informed decisions and delivering aid to the right place, quickly" (MapAction, 2011).

MapAction works with volunteer GIS professionals with expertise in disaster response and who are able to deploy to disaster management sites around the world, 365 days a year. MapAction provides maps and other geographical analytical products to various United Nations and national entities requiring maps to inform decision making.

The following are additional important international GIS for disaster management NGOs.

Humanitarian OpenStreetMap Team (HOT)

In their own words, the "the Humanitarian OpenStreetMap Team applies the principles of open source and open data sharing for humanitarian response and economic development" (Humanitarian OpenStreetMap Team, n.d.). Related to OpenStreetMap discussed in Chapter 3, HOT is has been very active in recent years with deploying volunteers to help create reference maps for countries lacking reference data. Notable projects in this regard include the 2010 Haiti earthquake where HOT began tracing roads and other features to supplement existing (but limited) reference data from high-resolution imagery that was collected within the first 48 hours after the disaster (Humanitarian OpenStreetMap Team, n.d.).

Crisis Mappers

Crisis Mappers is an international community with a strong, although not exclusive, emphasis on international humanitarian situations. You first learned about Crisis Mappers in Chapter 1 and the interview with Dr. Jennifer Zimeke. The key distinction of the Crisis Mappers from other "traditional" GIS and mapping entities for disaster management is their emphasis on utilizing new forms of mapping and other technology and data streams such as social media and crowdsourced data for disaster early warning and response (Ziemke, 2012). In their own words (Meier, 2014):

> Crisis Mappers leverage mobile & web-based applications, participatory maps, & crowd-sourced event data, aerial & satellite imagery, geospatial platforms, advanced visualization, live simulation, and computational & statistical models to power effective early warning for rapid response to complex humanitarian emergencies. As information scientists we also attempt to extract meaning from mass volumes of real-time data exhaust.

Ushahidi technology that you learned about in Chapter 1 is a good example of the mapping technologies used by the Crisis Mapper community. Additionally, the Crisis Mapper community is an excellent resource for keeping track of cutting-edge technology research and innovation with mapping and technology for disaster management and being apprised of worldwide humanitarian and disaster activities through a very active email group that anyone can join (see link in the Resources section).

Also as you learned in Chapter 1, the Crisis Mappers have gained a lot of attention for their ability to fill information gaps in disaster and humanitarian situations where there is limited media or other information access such as the situation in Syria. In this regard, there is the Standing By Task Force (SBTF), which is closely related to the broader Crisis Mapper community. The SBTF is a structure for organizing volunteers around the world within a virtual platform who can lend their varied expertise in different areas during a crisis for tasks such as crowdsourced event mapping, data analysis, and any other relevant skills and volunteers that the group can offer. For more information on the SBTF or to get involved, see http://blog.standbytaskforce.com/.

GISCorps

Established in 2003, the GISCorps is a group of volunteer GIS professionals organized by the Urban and Regional Information Systems Association (URISA). In their own words (URISA, 2009):

> GISCorps coordinates short-term, volunteer-based GIS services to underprivileged communities.
>
> GISCorps volunteers' services will help to improve the quality of life by:
>
> - Supporting humanitarian relief.
> - Enhancing environmental analysis.
> - Encouraging/fostering economic development.
> - Supporting community planning and development.
> - Strengthening local capacity by adopting and using information technology.
> - Supporting health and education related activities.
>
> GISCorps implements URISA's vision of advancing the effective use of spatial information technologies.
>
> GISCorps makes available highly specialized GIS expertise to improve the well being of developing and transitional communities without exploitation or regard for profit.
>
> GISCorps coordinates the open exchange of volunteer GIS expertise cooperatively among and along with other agencies.
>
> GISCorps strengthens the host community's spatial data infrastructure through implementation of the best and most widely accepted GIS practices.
>
> GISCorps foster development of professional organizations in host communities to help sustain and grow local spatial expertise.

Thus, GISCorps is not exclusively focused on disaster management activities, although this is an important service they do provide in a volunteer capacity with the important distinction from Crisis Mappers that they are primarily comprised of working, GIS professionals.

International Disaster Management Support Mechanisms

The following sections discuss international disaster management support mechanisms to give you a sense of the types of systems that exist to support GIS and related data. These mechanisms are important to know about as they often provide critical data and services to the international community that otherwise might not be available.

International Charter on Space and Major Disasters

The International Charter on Space and Major Disasters (or the International Charter) "aims at providing a unified system of space data acquisition and delivery to those affected by natural or man-made disasters through Authorized Users. Each member agency has committed resources to support the provisions of the Charter and thus is helping to mitigate the effects of disasters on human life and property" (The International Charter: Space and Major Disasters, 2014). Members of the International Charter include national space and other related agencies of countries worldwide such as the USGS in the United States, DLR in Germany, the European Space Agency (ESA), and the Chinese National Space Administration among others.

The basic process by which the charter works is as follows. When a disaster occurs, an authorized charter user (such as a relevant United Nations organization) can request activation of the charter from one of the charter members to provide relevant space-based information about the disaster area, for example, acquiring imagery through the USGS. If the request is approved, a project manager is assigned to handle the acquisition, processing, and handling of satellite-based assets. Furthermore, the project manager creates or will coordinate with a value-added reseller final products that are disseminated back to the requesting charter user. Often, these products are in the form of maps that show satellite imagery of impacted disaster areas (see link in the "References" section for examples). The International Charter is an important mechanism for providing satellite-based information for disaster areas that are either so large they require multiple countries to coordinate a disaster response or in developing countries that do not have a national space administration or the general capacity to acquire satellite-based imagery for their country.

Global Disaster Alert and Coordination System (GDACS)
The GDACS is "a cooperation framework between the United Nations, the European Commission and disaster managers worldwide to improve alerts, information exchange and coordination in the first phase after major sudden-onset disasters" (Global Disaster Alert and Coordination System, 2014). Thus, although the focus of GDACS is not exclusively on GIS and mapping, maps and mapping are an important component of GDACS-provided services. In particular, GDACS provides the members-only virtual On-Site Operations Coordination Centre (OSOCC) which is an online, virtual collaboration platform used by the United Nations and other humanitarian disaster response actors to coordinate and collaborate. Additionally, GDACS provides data, maps, and satellite imagery for the various events they track through the virtual OSOCC, the International Charter, and other mechanisms (see http://portal.gdacs.org/data).

World Bank GFDRR
In their own words, "the Global Facility for Disaster Reduction and Recovery (GFDRR) is a partnership of 41 countries and 8 international organizations committed to helping developing countries reduce their vulnerability to natural hazards and adapt to climate change. The partnership's mission is to mainstream disaster risk reduction (DRR) and climate change adaptation (CCA) in country development strategies by supporting a country-led and managed implementation of the Hyogo Framework for Action (HFA)" (Global Facility for Disaster Reduction and Recovery, 2014).

An initiative started by the World Bank, a key aspect of GFDRR is the emphasis on risk reduction, or mitigation, as defined in this chapter, and not disaster response. Furthermore, GFDRR works closely with developing countries for postdisaster recovery and redevelopment (World Bank, 2011). Of particular note in terms of GFDRR and GIS are the emphasis on open data initiatives for disaster risk reduction such as the Open Data for Resilience Initiative (https://www.gfdrr.org/opendri) and the broader World Bank open data initiative (first discussed in Chapter 3) where the World Bank continues to provide open and free access to country-level and subnational indicators for developing countries such as population characteristics, economic characteristics, industry characteristics, and other essential indicators for understanding the nature of vulnerability, risk, and resiliency for disaster risk reduction planning.

United Nations

The final chapter section discusses various United Nations organizations that are primarily involved in disaster management and also utilize GIS or related technology. The United Nations and disaster management is a another vast topic that could require an entire book unto itself as many United Nations organizations are involved in some aspect of disaster management, humanitarian relief, or related activities.

Office for the Coordination of Humanitarian Affairs: ReliefWeb

ReliefWeb (http://www.reliefweb.int/) is a United Nations–funded organization that is a suborganization of the UN's Office for the Coordination of Humanitarian Affairs (OCHA; http://ochaonline.un.org/). Since it is part of OCHA, the primary mission of ReliefWeb is to serve an information management coordination role through the collection, maintenance, and dissemination of humanitarian information to the humanitarian community.

A casual perusal of the ReliefWeb site reveals the vast amounts and categories of information that ReliefWeb provides. In order to better provide timely and relevant information to members of the humanitarian community, ReliefWeb offers several categories of information in web service and streaming formats such as Really Simple Syndication (RSS), YouTube, Facebook, and Twitter. Typically, the information offered is dynamic in terms of regular updates that are made to the information such as headlines, OCHA situation updates, job vacancies, map postings, and training.

UN-SPIDER

The United Nations Platform for Space-Based Information for Disaster Management and Emergency Response (UN-SPIDER) is the officially mandated UN program with a mission focused on building capacity for the use of space-based information within the full disaster management cycle (Backhaus et al., 2010). The mission of UN-SPIDER translates in tangible practice through its activities as (1) a *bridge* to connect space and disaster management communities as demonstrated in numerous international workshops that draw diverse participants from governments, NGOs, and academia; (2) a *facilitator* for institutional strengthening and capacity building as evidenced by numerous technical advisory missions to developing nations; and (3) a *gateway* to space information to support disaster management, as reflected in the UN-SPIDER Knowledge Portal (Tomaszewski, 2010).

The UN-SPIDER Knowledge Portal (KP; http://www.un-spider.org/) is a publically available, web-based gateway to collect and disseminate varied forms of knowledge and information relevant to the disaster management and satellite technology communities (Epler and Stumpf, 2011). Categories of knowledge and information available through the KP include, but are not limited to, the following (Epler and Stumpf, 2011):

- UN-SPIDER activity updates,
- space technology and information to support active disaster management activities,
- Technical Advisory Mission and workshop reports, and
- social networking forums to facilitate discussions related to space-based information quality, availability, accessibility, and costs.

The following interview provides further perspectives on UN-SPIDER.

UN-SPIDER Perspectives: Interview with Antje Hecheltjen*

Antje Hecheltjen (Figure 4.9) is Junior Professional Officer (JPO) in the position of Associate Programme Officer at UN-SPIDER in Bonn, Germany. She is responsible for GIS and remote sensing knowledge management and technical advisory activities. Before starting with UN-SPIDER in 2012, she was a research associate at the University of Bonn where she was involved in a variety of remote sensing–related research projects such as developing algorithms to automatically classify satellite images. Her specialization was change detection and working with *multitemporal* datasets as well as *multisensor* datasets. The following is the first of a two-part interview conducted for this book with Ms. Hecheltjen in January 2014. In this portion of the interview, she answers questions about her work with UN-SPIDER and the role of UN-SPIDER in broader international disaster risk management and emergency response activities. The second half of this interview is presented in Chapter 9 where Ms. Hecheltjen provides advice on getting a job in the GIS for disaster management field and discusses the future of GIS for disaster management at the international scale.

What types of disaster management activities does UN-SPIDER generally do?
There are three things we do. The first is our technical advisor support for institutional strengthening, which includes technical advisory missions to countries and follow-up training activities. During these missions we assess how countries are currently using space-based information for their disaster management activities. We try to identify any gaps, anything that can be improved, and submit to the government a detailed report with recommendations. We then, for example, follow up on these recommendations with training courses.

The second pillar of activities is knowledge management. This is mainly our online knowledge portal,[†] but not exclusively. The aim of the portal is to facilitate the access to data and to information for both the space community and for the disaster management community. The overarching goal is that we try to foster knowledge transfer. Let me give you an example. Currently, we are working together with our regional support offices[‡] to develop so-called recommended practices.

Figure 4.9   Antje Hecheltjen.

---

* The views expressed herein are those of the author(s) and do not necessarily reflect the views of the United Nations.
† UN-SPIDER, http://www.un-spider.org/.
‡ UN-SPIDER, "Regional Support Offices," http://www.un-spider.org/network/regional-support-offices.

More specifically, our regional support offices are developing step-by-step guides based on their own experience on how to find and access data on the web and how to process them, for example, on drought monitoring or flood management. Once ready, these practices will be disseminated through the knowledge portal.

The third pillar of our activities is the exchange of knowledge during expert meetings and workshops that we organize, which is also one of our channels for capacity building. During these events, the different actors that we [are] trying to reach can physically sit at the same table and discuss their needs and solutions. We have seen some great synergies developing from our events.

What I think is unique about UN-SPIDER is that we bring those communities together. There are not many platforms on which the space community and the disaster management community can interact. The space community has great tools, a lot of potential, but sometimes they have no awareness of what is really needed in developing countries. On the other side, the disaster management community often doesn't know how to use the information derived from space-based data.

Anything more specific you can talk about in terms of UN-SPIDER contributing to international disaster management in terms of geographic information or GIS technology?

We don't really generate information from satellite data ourselves. Our role is to enable people and institutions to do it themselves. One of the most important aspects of this is access to data. So, for example, in emergency situations, emergency responders need to know that there are mechanisms like the International Charter Space and Major Disasters or the Copernicus Emergency Management Service for emergency response* that provide and process satellite information for free. It is our role to raise awareness on these mechanisms as well as how to trigger them and who to contact. Instead of duplicating efforts of these existing mechanisms, we think it makes much more sense to make the right links.

It is only in the context of our training courses, that we actually work with satellite data and define datasets for demonstration purposes. Usually we use datasets of the country itself because it's more relevant for the participants; of course, we define the GIS and remote sensing methodologies that we want to teach them in very close cooperation with our partners. I would like to stress that without our regional support officers and other partners, we wouldn't be able to do as much as we do at the moment.

How did UN-SPIDER get involved in the Philippines disaster of 2013?

Especially our Beijing office was heavily involved in making the right connections in the Asian region to obtain relevant data on Typhoon Haiyan. We also set up an emergency support page on the knowledge portal. On this page we compiled all available datasets including baseline data and crowdsourced data for the public to freely access and updated it continuously. We wanted to offer a "one-stop shop," as it is very time-consuming to mine the web in order to find out who did which map, which data are available overall, and if any baseline data is available for this particular area. Similarly to the page we had set up for the 2010

---

* Copernicus, http://www.copernicus.eu/pages-principales/services/emergency-management/.

Haiti earthquake, our page for Typhoon Haiyan got lots of hits during that time, because it was the only page that offered everything in one place.

UN-SPIDER, covers all disasters, correct? Not just the big ones that get the media's attention. UN-SPIDER is there for the smaller disasters that don't get as much global attention, correct?

Yes. However, we only get involved in emergency situations if the respective country officially requests our support. Then it doesn't matter how big the disaster is. Mechanisms like the International Charter Space and Major Disasters have specific definitions of what a major disaster is and may not accept requests on events that do not fulfill these requirements.

What types of specific GIS or geographic information things do you do?

Here in Bonn, where I am stationed, the main task is knowledge management. Mainly we're operating our knowledge portal, which needs a lot of web research and content production regarding the use of space-based information for disaster management. For some specific topics, my technical background is quite helpful. For example, one of my tasks at the moment is to build up repositories on available datasets such as satellite imagery and baseline data. Another repository compiles available software such as remote sensing and GIS software. There's also the idea to establish another database on tutorials and methods so that our users can learn how to use satellite data to generate useful maps and other products. That's in the end what we want to do; we want to enable practitioners to use the information that's out there in a useful way in their countries.

Another one of my tasks is technical advisory support. Last year, for example, I organized and supported a training course in the Dominican Republic* where we trained an interdisciplinary team from different ministries, university, military to use satellite imagery for flood management. Together with our regional support offices in Latin America, CATHALAC,[†] and IGAC,[‡] we conducted this training and gave an introduction to remote sensing and GIS for the particular context of floods.

As a conclusion to this chapter section, the following interview discusses broader perspectives on GIS for Disaster Management within the United Nations.

GIS, Disaster Management, and the United Nations: Interview with Dr. Jörg Szarzynski[§]

Dr. Szarzynski (Figure 4.10) is Education Programme Director and head of the Enhancing Graduate Educational Capacities for Human Security (EGECHS) section at the United Nations University Institute for Environment and Human Security (UNU-EHS) in Bonn, Germany. He holds an MSc in geography and a PhD in Physical geography and atmospheric sciences with key competencies in climatology, geobiophysics, remote sensing, and tropical ecology. He is the responsible program officer for investigating the teaching–research nexus to enhance

---

[*] UN-SPIDER, "Dominican Republic: Training on Space-Based Information for Floods," http://www.un-spider. org/about-us/news/en/6655/2013-05-14t165000/dominican-republic-remote-sensing-training-inaugurated.

[†] UN-SPIDER, http://www.un-spider.org/network/regional-support-offices/water-center-humid-tropics-latin-america-and-caribbean-cathalac.

[‡] UN-SPIDER, http://www.un-spider.org/network/regional-support-offices/colombia-regional-support-office

[§] The views expressed herein are those of the author(s) and do not necessarily reflect the views of the United Nations.

Figure 4.10   Dr. Jörg Szarzynski.

UNU-EHS graduate education and research training. His duties comprise the management of the Joint Master Program titled "Geography of Environmental Risks and Human Security" (a joint master program with the University of Bonn), as well as the UNU-EHS PhD program and training courses for postgraduate professionals. In addition, he actively organizes international seminars, workshops, lectures, and training courses focused on remote sensing and GIS applications for disaster risk reduction, contributing to the development of training materials and e-learning modules and the supervision of MSc and PhD scholars. His research interests include environmental change and sustainable development research, early warning systems and disaster management, and intercultural training and education concepts.

The following is the first of a two-part interview conducted for this book with Dr. Szarzynski in July 2013. In this portion of the interview, he answers questions concerning the general role of the United Nations in broader international disaster management activities. The second half of this interview is presented in Chapter 9 where Dr. Szarzynski provides advice on developing a career in international disaster management with the United Nations and GIS.

What types of disaster management activities does the UN get involved in?
When we talk about international disaster management activities and the role of the UN, I think in the first instance we need to mention that, at the international level the United Nations, and specifically UN OCHA (the UN Office for the Coordination of Humanitarian Affairs), take the coordinating role whenever a disaster occurs and the affected country is asking for international support; this is very important due to the sovereignty of a state in respect to civil protection. In such a case, UN OCHA may take over the role to assist in the coordination of incoming international relief at national level. This role is also confirmed and accepted by the European Union Community mechanism for civil protection, just to mention another big player at the international level. Furthermore, OCHA's role is always to act as a link between

international responders and the government of the affected country, together with national and local authorities such as national civil protection units or national disaster management organizations. OCHA staff members have to make sure that all the efforts coming from the international level are somehow coordinated. That means relief support stemming from other players from the United Nations, for example, the World Food Program (WFP),* the World Health Organization (WHO),† UNDP (UN Development Programme),‡ UNHCR (UN High Commission for Refugees),§ and other mandated organizations within the United Nations family is coordinated on-site. Structures like the OSOCC (On-site Operations Coordination Center) are used to help local authorities in a disaster-affected country to coordinate international relief in the most effective way.

Are there any other UN groups involved in disaster management?

Well, talking about UN OCHA, we also have to mention the so-called UNDAC (United Nations Disaster Assessment and Coordination¶) system, which is managed by the Emergency Services Branch of OCHA based in Geneva. UNDAC teams are deployed to support the UN and governments of disaster-affected countries during the first phase of a rapid-onset emergency. These teams are usually the first people on the ground after a disaster took place and their specific task is to assist in the coordination of incoming international relief at the site of the emergency and to conduct first-needs assessments. This includes, for example, information on what is needed first such as food, shelter, medical support, or what else might have highest priority. Secondly, the so-called UN cluster approach needs to be mentioned. Currently, there are 11 clusters consisting of different groups of humanitarian organizations, from the UN but also non-UN agencies, working together within the major sectors of humanitarian action, e.g., food, shelter, and health, but also logistics, emergency telecommunications, or education. WFP comes in, UNHCR [UN High Commission on Refugees], WHO [the World Health Organization], and UNICEF** [UN Children's Fund] especially in the case of large-scale disasters where huge parts of the population have been effected. In terms of internally displaced people (IDPs) and refuges, UNHCR and UNICEF play a leading role to make sure that human rights are taken care of. WHO, for instance, tries to control cascading effects, such as the high danger of epidemic diseases in the aftermath of a flood event. In almost all the countries worldwide, there is most likely a UNDP office and very often the leading officer takes the role of resident coordinator. Frequently, this is the senior-most UN officer in the country who automatically takes over the first coordinating steps in a disaster response as the humanitarian coordinator. When we look more carefully at disaster risk reduction, we also have to refer to UNISDR,†† the United Nations International Strategy for Disaster Reduction (discussed further

---

* UN World Food Programme, http://www.wfp.org/.
† World Health Organization, http://www.who.int/.
‡ UN Development Programme, http://www.undp.org/.
§ UN High Commission on Refugees, http://www.unhcr.org/.
¶ UN OCHA, http://www.unocha.org/what-we-do/coordination-tools/undac/overview
**UNICEF, http://www.unicef.org/.
††UNISDR, http://www.unisdr.org/.

in Chapter 8). Prevention and preventive measures, as well as the coordination of disaster reduction synergies among disaster reduction activities, are certainly some of the major goals on the agenda of this UN agency.

How extensive do you think the use of GIS is in UN disaster management activities and do you think that the current level of GIS use is sufficient or where do you see room for improvement?

The professional handling of GIS and RS applications within international disaster management activities requires experienced, professional organizations, such as the Center for Satellite-Based Crisis Information (ZKI) from the German Aerospace Center (DLR),* SERVIR,[†] or UNITAR/UNOSAT,[‡] For example, as so-called value adders, they conduct rapid mapping procedures on behalf of the International Charter Space and Major Disasters. As time and accuracy, but also rapid access to data from satellites are most crucial factors, this is the level where really professional RS/GIS handling and global collaboration are necessary. Active field support somewhere in the world, for instance, to do rapid assessment in the area affected by a disaster, is carried out by organizations such as the UK-based MapAction. In all of these cases, I would like to underline, that especially GIS carries the function to improve the awareness and especially the resilience towards disaster, and also the disaster response preparedness. Thus, this kind of technology can be used at more simple levels to expand the spatial thinking capacity of people. Take the example of an evacuation in a very time-crucial situation like a tsunami event, when it is really of paramount importance that, first of all, people have to be aware of the situation, people also have to be trained, so that automatically, when a siren starts on a beach that they exactly know what to do and in which direction they should evacuate. And here, I do believe, a lot of awareness raising is still necessary including a tremendous role that Geographic Information Systems may play in this context.

As a disaster risk reduction educator, what opportunities and challenges do you see with people learning GIS in an international disaster context?

It is clear that, on the one side, we need the real "techies," I mean professional people who know how to handle a comprehensive GIS system. This usually takes you quite some time to really learn most functionalities of, let's say ArcGIS from Esri. On the other hand, for a lot of people, it is sufficient just to understand a little bit of how you can use GIS in the best way, just to increase your spatial thinking and spatial awareness of a situation. However, to adequately support an international large-scale disaster appropriately, you will need a core team of professional GIS users. The group of people knowing a little bit on GIS, or how to handle spatial data, opens GIS to a broad range of new opportunities. If these people can be trained in the general and basic handling of such systems together with GPS and other technologies combined with some Google applications, I think this will generally increase the recognition of spatial thinking and also the potential output and contribution this community may deliver in terms of crowdsourcing, filtering crisis-relevant, geo-referenced information stemming from social networks, etc.

---

* ZKI, http://www.zki.dlr.de/.

[†] NASA SERVIR, http://www.nasa.gov/mission_pages/servir/#.UvLl4PldWSo.

[‡] UNITAR, http://www.unitar.org/unosat/.

Can you discuss some of your own specific activities related to disaster risk reduction and GIS education?

Personally, I started my professional career as an environmental scientist where I used GIS and remote sensing in frame of typical applications, such as determination of land-use, land-cover changes, to gain a better understanding of spatial patterns, especially mechanisms shaping current changes in biodiversity, or integrating bio-physical and socioeconomic data. Later on, when I was involved in the German-Indonesian Tsunami Early Warning System—GITEWS—the use of sensor web technologies in favor of disaster management and the generation of early warning relevant information became a much more prominent role. Finally, during my time working with the UN-SPIDER platform, remote sensing and GIS technologies in support of risk and disaster management were in the center of our daily work. Since UN-SPIDER is acting as an information broker and a gateway between the space- and the disaster-management community it was always our duty to describe these technologies for nontechnical persons in an appropriate way and to provide access to related data and information whenever needed in the aftermath of a disaster.

In my current position here at the United Nations University, Institute for Environment and Human Security, it is the educational aspect that came to the forefront. During our training courses here in Bonn, but also in South Africa, Togo, and elsewhere, we usually try to take the perspective of stakeholders, who are not necessarily from the technological community. Reset yourself to "point zero," and then start to look at these very sophisticated GIS packages currently available on the market, and then try to ask yourself: what is really essential for a disaster manager in the field? GIS professionals have gained their knowledge and expertise by studying GIS and remote sensing in geography, geodesy, cartography, geophysics, or other comparable disciplines. In our training courses we try to educate and qualify academics and practitioners accordingly to the growing request and demand for getting some profound insights into geospatial technologies based on tailor-made modules adapted for professionals who don't have the time to spend weeks or months at a university. Those types of people are looking for either some training that they can do on the base of distance learning, e-learning, or perhaps a customized module of perhaps two to three weeks where they can really learn more than just an overview. You can take them to some hands-on exercises where they learn the basics, where they can see the potential and also evaluate the potential of this kind of technology for themselves. And, of course, they also learn what is out there in the web in terms of open resource software and also simpler technologies that they can use in their daily business. So, I think we have to differentiate between the real professionals and those who just can make use of this certainly very useful technology within so many different areas of professional worker.

Please let me add one final point. We have meanwhile discussed a lot about GIS for disaster management, but just think on all the other different activities that take place within the UN, such as the United Nations Environment Programme (UNEP)* and its focus on environmental problems and topics. GIS and remote

---

* UNEP, http://www.unep.org/.

sensing already played a very crucial role for UNEP for a long time. Another area of application comes with the topics of public health. Again, remote sensing and GIS are becoming increasingly important, for instance, to better understand different factors steering environmental changes and the corresponding influences on changes in the distribution of disease patterns. Telemedicine, to mention a further growing field of application, is more or less fully related to satellite-based telecommunication. This technology opens a new cosmos of possibilities how people can communicate and exchange information and transfer knowledge all over the world. And a necessary medical treatment in a remote area of the world, let's assume an urgent surgery on a battleship in the middle of the Pacific Ocean, can be supported via satellite-based telecommunication and combined video channel by a colleague, located elsewhere in the world in a modern clinic environment. Basically, satellite-based technologies open a lot of options for numerous activities that the UN is carrying out. For example, if we talk about the work related to migratory animals done by our UN colleagues from the convention on migrating species, they also use remote sensing and other sensor tracking methods in order to observe and monitor all the pathways of migrating animals to provide a better understanding of diversity patterns worldwide.

Back to the field of natural hazards and disasters, just look at the example of the terrifying Great Eastern Japan earthquake in March 2011 where we were confronted with this shocking example of a cascading disaster. Initially, there was a strong earthquake, subsequently followed by a devastating tsunami, and finally ending up with the nuclear incident contaminating larger areas around Fukushima and along the eastern coast of Japan. All of these different disaster types need to be tackled in a very specific way, but as one denominator, remote sensing and GIS was used by a lot of agencies just to describe their very own specific requests, starting with the first assessment of what has happened, followed by a more detailed damage assessment, specifically for search and rescue teams and so on. And finally, also the work of our colleagues from the International Atomic Energy Agency in Vienna was supported by information generated through remotely sensed data and GIS analysis. As one example, when they overlaid meteorological information and the distribution patterns of contaminated particles over an area, buffer zone calculations from GIS was used to track these pathways.

CHAPTER SUMMARY

In this chapter, you learned about the specific relationships between disaster management and Geographic Information Systems. You first learned about distinctions among terms such as *emergency*, *disaster*, *crisis*, and *catastrophe*. You were then formally introduced to the concept of the disaster management cycle and its four phases of response, recovery, mitigation, and planning. The chapter then discussed the role of GIS within disaster management policy and practice from the perspective of the United States. Specifically, you learned about NIMS and how geospatial information is formally considered in this national policy for planning and communications and information management. You also learned about the ICS, which is an incident management framework designed for all hazard response, and again, how GIS plays an important role in planning activities. In this discussion, you also learned how the role of GIS technical specialists is becoming more formally established and thus has the potential for greater employment opportunities in the GIS for disaster management field. You were also briefly introduced to the United States DHS GeoCONOPS and the United States NSDI. The chapter then discussed the use of GIS at different levels of government in the United States. Specifically, you learned about the use of GIS in cities, towns, and counties, the use of GIS at the state level, GIS clearinghouses, and finally, the use of GIS at the national level through organizations such as FEMA. This chapter section also discussed other US federal agencies that use GIS for disaster and humanitarian assistance at the international scale. You also got some perspective on how the private sector relates with government agencies and disaster management practice.

The second half of the chapter discussed the international disaster management community and GIS. Specifically, you learned about four NGOs that have an explicit connection with GIS for disaster management—MapAction, the Humanitarian OpenStreetMap Team (HOT), the Crisis Mappers, and the GISCorps. Next, you learned about international disaster management support mechanisms. The purpose of presenting these items was to give you a sense of how the world responds (in terms of GIS and related technology) when disasters occur and about global efforts to reduce the risks associated with disasters from things such as climate change. Specifically, you learned about the International Charter on Space and Major Disasters, which is used to provide satellite imagery for large-scale disaster situations or for developing countries that do not have space-based imagery capabilities and capacities. You also learned about GDACS, which is a mechanism for automated disaster event detection, analysis, and coordination of humanitarian actors through virtual platforms. Finally, you learned about the World Bank's Global Disaster Risk Reduction program and its open data initiatives that are a topic of particular interest in terms of disaster risk reduction. The chapter ended with a discussion of several United Nations organizations involved with disaster management, GIS, and general information management needs during global disaster response.

Throughout the chapter, you heard perspectives on the topic of GIS for disaster management from several working professionals from many of the specific organizations discussed. You should read these interviews carefully and keep the perspectives and advice they offered in mind as you move forward in your own career in GIS for disaster management and depending on where your interests lie in terms of where you might want to work ranging from local government all the way to the United Nations.

The next four chapters delve deeper into the use of GIS for disaster management for each specific disaster cycle phase to which you were introduced in this chapter to give you specific ideas, techniques, and advice and for your own GIS for disaster management activities.

DISCUSSION QUESTIONS AND ACTIVITIES

1. Download a copy of NIMS from http://www.fema.gov/pdf/emergency/nims/NIMS_core.pdf (or whatever the current link is). What other aspects of NIMS do you think are spatial in nature and would lend themselves to the use of GIS—even if the words "geospatial" or "GIS" are not actually used?

2. Go on the Internet and look for your local government's GIS division. Compare your local GIS division (if it exists) with a neighboring town or your county or state. What sorts of data and services do each of these provide? From what you find, how prepared do you think your local governments is for GIS and disaster management activities?

3. Look through the GIS jobs clearinghouse website cited earlier in the chapter. Examine some of the jobs listed. How well you think you are prepared to get a job in GIS and what additional skills do you think you would need? If you find any jobs with an explicit disaster management or related focus, what other sorts of training might you need? Do a general search for "GIS and disaster management jobs" (or vary the term *disaster* with the other terms such as *emergency* or *crisis*). What sorts of jobs do you find, and how well prepared and/or qualified do you think you are for them?

4. Look through many of the case studies provided on the MapAction or GISCorps websites. What GIS for mapping activities do you find particularly interesting and why?

5. Look through the many charter activations on the International Charter on Space and Major Disasters website. How useful do you think the maps provided are? When looking at the maps, make particular note of when the actual disaster occurred and when the final map products were delivered. Do you think there is too much of a time gap between the two?

6. What other United Nations organizations can you find that do activities related to disaster management? If not explicitly mentioned, how might the activities that you do find be related to GIS or might incorporate GIS (remember—think spatially!)?

RESOURCES NOTES

For further reading on emergency management, see Phillips, B.D., D.M. Neal, and G. R. Webb. 2012. *Introduction to Emergency Management*. Boca Raton, FL: CRC Press.

Examples of FEMA GIS Specialist positions as of 2013:

GIS Analyst: http://www.fema.gov/media-library-data/20130726-1918-25045-5795/gisanalyst.pdf

GIS Supervisor: http://www.fema.gov/media-library-data/20130726-1918-25045-9979/gissupervisor.pdf

GIS Field Data Entry Technician: http://www.fema.gov/media-library-data/20130726-1918-25045-4912/gisfielddataentrytechnican.pdf

For more information on GeoCONOPS in practice, see "IS-62: GeoCONOPS In-Practice: Homeland Security Geospatial Concept-of-Operations (GeoCONOPS)," FEMA Emergency Management Institute, https://training.fema.gov/EMIWeb/IS/courseOverview.aspx?code=IS-62.

For a list of other international development agencies, see Wikipedia, http://en.wikipedia.org/wiki/List_of_development_aid_agencies.

For more information on the USAID GeoCenter, see the ArcGIS.com website, http://www.arcgis.com/home/group.html?owner=usaidgeocenter&title=USAID%20 GeoCenter.

For examples of other HOT projects, see the OpenStreetMap website, http://hot. openstreetmap.org/projects.

The Crisis Mappers Google Group and email list can be found at: http://groups. google.com/group/crisismappers?hl=en.

For information on GISCorps a list of its projects, see http://www.giscorps.org/ index.php?option=com_content&task=view&id=22&Itemid=59.

For a full list of International Charter on Space and Major Disasters members, see http://www.disasterscharter.org/web/charter/members.

For a full list of International Charter on Space and Major Disasters activations and associated products, see http://www.disasterscharter.org/web/charter/activations.

REFERENCES

Alexander, David. 2013. *Homeland Security Geospatial Strategy*, Presentation Soundtrack Esri Fed UC 2014, National Security Immersion Summit, Washington, DC.

ArcGIS. 2014. "Hurricane Sandy: The NYC story," ArcGIS.com, February 7, http://www.arcgis. com/home/item.html?id=05cf92323437411c8fbba4e774d8f8b3 (accessed April 5, 2014).

Backhaus, Robert, Lorant Czaran, Natalie Epler, Michael Leitgab, Young Suc Lyu, Shirish Ravan, David Stevens, Peter Stumpf, Joerg Szarzynski, and Juan-Carlos Villagran de Leon. 2010. "Support from space: The United Nations Platform for Space-Based Information for Disaster Management and Emergency Response (UN-SPIDER)." In *Geoinformation for Disaster and Risk Management: Examples and Best Practices*, edited by O. Altan, R. Backhaus, P. Boccardo and S. Zlatanova. Copenhagen, Denmark: Joint Board of Geospatial Information Societies.

Bernard, Lars, Ioannis Kanellopoulos, Alessandro Annoni, and Paul Smits. 2005. "The European geoportal: One step towards the establishment of a European Spatial Data Infrastructure." *Computers, Environment and Urban Systems* 29 (1):15–31.

Cutter, Susan L., and Christopher T. Emrich. 2006. "Moral hazard, social catastrophe: The changing face of vulnerability along the hurricane coasts." *The ANNALS of the American Academy of Political and Social Science* 604 (1):102.

Epler, Natalie, and Peter Stumpf. 2011. "The UN-SPIDER knowledge management and information dissemination system." In *Proceedings of the Fourth United Nations International UN-SPIDER Bonn Workshop on Disaster Management and Space Technology: The 4C – Challenge: Communication – Coordination – Cooperation – Capacity Development*, edited by A. Froehlich. Bonn, Germany: Deutschen Zentrums für Luft- und Raumfahrt (DLR).

Federal Emergency Management Agency (FEMA). 2009. *IS-700.A: National Incident Management System (NIMS) An Introduction: Unit 4: NIMS Communications and Information Management*, FEMA, http://training.fema.gov/EMIWeb/is/IS700a/SM%20files/IS700A_StudentManual_ L4.pdf (accessed April 5, 2014).

Federal Emergency Management Agency (FEMA). 2013. "About FEMA," FEMA, http://www.fema. gov/about-fema (accessed April 3, 2014).

Federal Emergency Management Agency (FEMA). 2013. "Hurricane Sandy: Timeline," FEMA, http://www.fema.gov/hurricane-sandy-timeline (accessed April 3, 2014).

Federal Geographic Data Committee (FGDC). 2007. "National Spatial Data Infrastructure," FGDC, http://www.fgdc.gov/nsdi/nsdi.html (accessed April 5, 2014).

Global Disaster Alert and Coordination System. 2014. GDACS website, http://www.gdacs.org/ (accessed February 8, 2014).

Global Facility for Disaster Reduction and Recovery. 2014. GFDRR website, https://www.gfdrr.org/ (accessed February 8, 2014).

Griffiths, David M. 2006. "Managing information: A practical guide," http://www.managing-information.org.uk (accessed January 15, 2014).

Humanitarian OpenStreetMap Team. n.d. "Haiti," n.d. Humanitarian OpenStreetMap Team, http://hot.openstreetmap.org/projects/haiti-2 (accessed April 5, 2014).

Humanitarian OpenStreetMap Team. n.d. Humanitarian OpenStreetMap Team website, http://hot.openstreetmap.org/ (accessed April 5, 2014).

Information Systems for Crisis Response and Management (ISCRAM). 2014. "International Community on information systems for crisis response and management," ISCRAM, http://www.iscram.org/ (accessed February 7, 2014).

International Charter: Space and Major Disasters. 2014. Disaster Charter Homepage, http://www.disasterscharter.org/home (accessed February 8, 2014).

International Federation of Red Cross and Red Crescent Societies. n.d. IFRC website, https://www.ifrc.org/en/what-we-do/disaster-management/about-disaster-management/ (accessed April 3, 2014).

Kevany, Michael J. 2003. "GIS in the World Trade Center attack: Trial by fire." *Computers, Environment and Urban Systems* 27 (6):571–583.

Kunreuther, H. 2002. "Risk analysis and risk management in an uncertain world." *Risk Analysis* 22 (4):655–664.

Lewis, Tanya. 2013. "The scientist who helped save New York's subway from Sandy," Live Science, October 28, http://www.livescience.com/40736-scientist-who-saved-new-york-subway-from-sandy.html (accessed April 5, 2014).

Lighthouse Readiness Group. 2012. "The difference between a crisis, emergency, and disaster," Lighthouse Readiness Group, http://lighthousereadiness.com/lrg/difference-crisis-emergency-disaster/ (accessed February 7, 2014).

Locker, Ray. 2013. "Military maps to focus on natural disasters, analysis," *USA Today*, November 19, http://www.usatoday.com/story/nation/2013/11/19/nga-geoanalytics-map-natural-disasters/3628911/.

MacEachren, Alan M. 2005. "Moving geovisualization toward support for group work." In *Exploring Geovisualization*, edited by J. Dykes, A. MacEachren and M. J. Kraak. New York: Elsevier.

MapAction. 2011. "What we do," MapAction, http://www.mapaction.org/about/about-us.html (accessed February 8, 2014).

MapAction. 2013. "Philippines, November 2013, MapAction, http://www.mapaction.org/deployments/depldetail/224.html (accessed April 3, 2014).

Masser, Ian. 2005. *GIS Worlds: Creating Spatial Data Infrastructures*. Vol. 338. Redlands, CA: ESRI Press.

Meier, Patrick. 2014. "CrisisMappers: The Humanitarian Technology Network," CrisisMappers website, http://crisismappers.net/ (accessed April 5, 2014).

Mileti, Dennis S. 1999. *Disaster by Design: A Reassessment of Natural Hazards in the United States*. Washington, DC: Joseph Henry Press.

Nourbakhsh, Illah, Randy Sargent, Anne Wright, Kathryn Cramer, Brian McClendon, and Michael Jones. 2006. "Mapping disaster zones." *Nature* 439 (7078):787–788.

NYC Information Technology and Communications. 2014. "Citywide IT services: GIS NYCityMap," NYC, http://www.nyc.gov/html/doitt/html/citywide/gis.shtml (accessed February 7, 2014).

Phillips, Brenda D., David M. Neal, and Gary R. Webb. 2012. *Introduction to Emergency Management*. Boca Raton, FL: CRC Press.

Quarantelli, E.L. 2006. "Catastrophes are different from disasters: Some implications for crisis planning and managing drawn from Katrina," SSRC, http://understandingkatrina.ssrc.org/Quarantelli/ (accessed February 7, 2014).

Tomaszewski, Brian. 2010. Gateway, Bridge and Facilitator. *GIM International*. (24)3, or see: http://www.gim-international.com/issues/articles/id1507-Gateway,_Bridge_and_Facilitator.html (accessed August 15, 2014).

United Nations Office for Disaster Risk Reduction (UNISDR). 2007. "Terminology: Disaster," UNISDR, http://www.unisdr.org/we/inform/terminology (accessed February 7, 2014).

United Nations Office for Disaster Risk Reduction (UNISDR). n.d. "What is disaster risk reduction?" http://www.unisdr.org/who-we-are/what-is-drr (accessed July 10, 2014).

United States Agency for International Development (USAID). 2011. "USAID launches new geocenter," USAID, November 10, http://www.usaid.gov/news-information/press-releases/usaid-launches-new-geocenter (accessed February 7, 2014).

United States Department of Homeland Security. 2008. "National Incident Management System," United States Department of Homeland Security, http://www.fema.gov/national-incident-management-system (accessed May 25, 2014).

United States Department of Homeland Security. 2013. *Homeland Security Geospatial Concept of Operations (GeoCONOPS) Quick Start Guide*, DHS, http://www.nsgic.org/public_resources/HLS_GeoCONOPS_QSG_v5.pdf (accessed May 25, 2014).

United States Department of Labor. n.d. *Incident Command System Technical Specialists: Geographic Information System (GIS) Specialist*, Occupational Safety and Health Administration, https://www.osha.gov/SLTC/etools/ics/tech_special.html#geo (accessed February 7, 2014).

United States Department of State. n.d. Humanitarian Information Unit website, https://hiu.state.gov/Pages/Home.aspx (accessed April 5, 2014).

Urban and Regional Information Systems Association (URISA). 2009. *Mission Statement*, GISCorps, http://www.giscorps.org/index.php?option=com_content&task=view&id=16&Itemid=52 (accessed February 8, 2014).

World Bank. 2011. "Geographic information system: A revolution by stealth," World Bank, http://web.worldbank.org/external/default/main?contentMDK=23035843&menuPK=6454478&pagePK=7278674&piPK=64911825&theSitePK=5929282#3 (accessed February 8, 2014).

Yasin, Rutrell. 2013. "NGA, geospatial community plan a clear picture of major disasters," GCN, http://gcn.com/Articles/2013/04/12/NGA-geospatial-community-cloud.aspx (accessed April 5, 2014).

Ziemke, Jen. 2012. "Crisis mapping: The construction of a new interdisciplinary field?" *Journal of Map & Geography Libraries: Advances in Geospatial Information, Collections & Archives* 8 (2):101–117.

5

Geographic Information Systems and Disaster Planning and Preparedness

CHAPTER OBJECTIVES

Upon chapter completion, readers should be able to

1. identify essential Geographic Information Systems (GIS) datasets that should be acquired, developed, and curated during disaster planning activities;
2. understand organizational perspectives in terms of memorandums of understanding and cooperation agreements on the use of GIS datasets and other resources for disaster management;
3. discern how GIS can be used as a support tool for common disaster planning and preparation activities, such as evacuation route and zone planning;
4. understand the importance of scenario modeling for training purposes in the use of GIS to help answer what-if questions for disaster planning;
5. understand how GIS can be used for public outreach and citizen participation during disaster planning and preparation activities; and
6. understand the nature of GIS and disaster management planning and preparation on an international scale.

INTRODUCTION

This chapter introduces the concept of Geographic Information Systems for disaster preparedness, and by extension, GIS for disaster planning given the close connection between preparedness and planning. This chapter is presented as the first chapter to focus on each disaster cycle phase, as discussed in Chapter 4. In fact, you can see this chapter as "preparing" you for the chapters that follow, much in the same way GIS must first be prepared for other disaster cycle phases. For example, when a disaster event occurs, it is not the time to meet to make plans and establish operations such as acquiring essential base data layers, conducting GIS training, formulating data-sharing agreements with other organizations, or running what-if scenarios. These types of activities must be done *before* an actual event occurs. Introducing new concepts, datasets, technologies, and ways of conducting disaster management activities during a disaster response can divert precious time, attention, and resources away from time-critical, pressing needs. Thus, it is essential that proper plans are in place before an event occurs.

The term *preparedness*, however, is itself somewhat problematic as it implies some measurable level that can be achieved. For example, stating that one is *well prepared*—what does that really mean? Preparedness levels are difficult to measure when one considers that disasters operate at multiple scales and the multitude of factors that influence preparedness such as culture and history (Phillips, Neal, and Webb, 2012). For example, people who live near in major river may be more prepared for the flooding event due to a long history and culture of dealing with floods as opposed to people who live in an urban environment and have never experienced the kind of flooding that was seen in 2012's Hurricane Sandy event in the New York City (NYC) region, which caught many citizens off guard (Plumer, 2012). This same idea extends to some of the concepts discussed in Chapter 4 in terms of culture and history of government, private sector, and other organizations that utilize GIS for disaster management. For example, the small county government with limited GIS capacity will not be prepared to handle a major disaster event if a history of such events are not part of past experiences and memory, as opposed to a large city or county government, which will be more prepared for a wider variety of disaster events and supporting disaster management of those events with GIS.

From a mapping and spatial thinking perspective, also consider how prepared or perhaps more accurately, underprepared, average citizens are in terms of using maps to make decisions and understand situations they might face. For example, in today's world, most people rely so heavily on their Global Positioning System (GPS) devices, they are unable to make routing decisions or other navigation tasks if GPS capabilities are not available to them. If power, the Internet, or even general use of phones, tablet computers, and other technology support mechanisms are lost, serious problems can occur because people rely too much on such technology and not being able to function without it. Challenges like these are part of broader issues around general citizen preparedness that are challenging to address due to the complex nature of citizen preparedness, such as making citizens aware of natural, technological, and terrorist threat characteristics as well as how to recover from a disaster (Federal Emergency Management Agency [FEMA], 2004). However, GIS can play its own part within this broader issue by helping citizens and governments understand the essential location and spatial aspects of disaster preparedness, which can then translate into practice during a disaster response.

The remaining chapter sections are written primarily from the perspective of a GIS professional or student of GIS and disaster management looking to utilize GIS to support disaster preparedness activities as well as prepare and plan GIS itself. Use of GIS by non-GIS professionals, such as private citizens, is beyond the scope of this chapter. However, the end of the chapter presents ideas for using GIS as a public outreach and citizen participation tool for disaster planning. The next chapter section discusses preparing GIS itself in terms of essential datasets that must be in place, and preparing the technology and processes that are used during other disaster management cycle phases.

TECHNOLOGY AND DATASET PLANNING AND PREPARATION

Essential Disaster Management Map Layers

Although disasters manifest themselves in a wide variety of forms based on numerous underlying hazards coupled with the idiosyncrasies of people, places, culture, and history, it is prudent to prepare and curate certain map data sets common to all disaster situation types. These map datasets can be considered *reference* data layers as per the conversations on different map types in Chapter 2. From these essential reference map layers, other layers particular to a geographic context can be included to address specific needs and disaster management functions. Remember too, that as you develop your own data resources through disaster planning activities, it is essential to maintain proper and updated metadata so that the data you create or acquire is easily shared and understood by others that may need to use it.

One useful source for considering essential disaster management map layers is the Geographic Information Framework Data Standard developed by the US Federal Geographic Data Commission (FGDC) that "establishes common requirements for data exchange for seven themes of geospatial data that are of critical importance to the National Spatial Data Infrastructure (NSDI), as they are fundamental to many different Geographic Information Systems (GIS) applications" (Federal Geographic Data Committee, 2008).

The seven themes, along with illustrative, representative graphics of what each might look like in GIS, are shown in Table 5.1.

Note, too, that many of the data themes outlined in Table 5.1 are available in common, integrated reference map sources such as Google Maps, Google Earth, or OpenStreetMap, and hence the popularity of these sources as they provide quick and rapid access to many of these data themes without any need for specialized GIS training. However, it is important to keep in mind that despite the ease of use and accessibility of sources like Google Maps, they often cannot be changed or tailored if specific needs arise. For example, although the Google base map may be quite sufficient for many needs, its cartographic styling cannot be modified, and this could cause potential problems when adding other, context-sensitive map layers on top of the Google base map. Thus, if the capacity exists, it is ideal to develop one's own reference data layers typically in a vector data model format that can be modified and altered as needed.

Additional Sources of Ideas for Essential Disaster Management Map Layers

As stated previously, the map layer themes previously discussed are meant to be a general starting point for incorporating other essential disaster management map layers. It is difficult if not impossible to predefine all of the map layers that would be needed for disaster management activities. Thus, the following are some sources you are encouraged to consider examining for ideas on other specific data layers to acquire as part of disaster management GIS planning activities. As you look through these, and other sources, keep the particulars of your geographic context in mind. For example, if you live in a small town, data layers such as subways and railroads may not be relevant and thus not necessary to acquire and to commit limited resources for data acquisition and curation. Also, it is important to clarify terms you were first introduced you in Chapter 3 where you learned about the *data model* concept. This is the idea of how geographic entities are represented in a computer such as the raster and vector data models. In the context of this chapter,

Table 5.1 FGDC Geographic Information Framework Data Standard Themes Relevant to Disaster Management Planning with Illustrative Examples

Theme	Description	Illustrative Example
Cadastral data	Geographic extent of the past, current, and future property rights and interests in real property including the spatial information necessary to describe that geographic extent.[a]	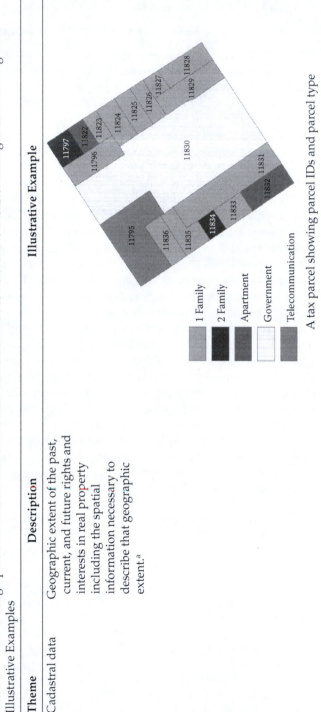 A tax parcel showing parcel IDs and parcel type

Digital orthoimagery

High-resolution aerial images that combine the visual attributes of an aerial photograph with the spatial accuracy and reliability of a planimetric map.[b]

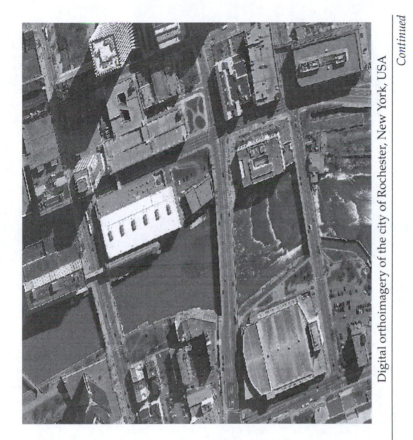

Digital orthoimagery of the city of Rochester, New York, USA

Continued

Table 5.1 (*Continued*) FGDC Geographic Information Framework Data Standard Themes Relevant to Disaster Management Planning with Illustrative Examples

Theme	Description	Illustrative Example
Elevation	Height above a specific vertical reference.[c]	

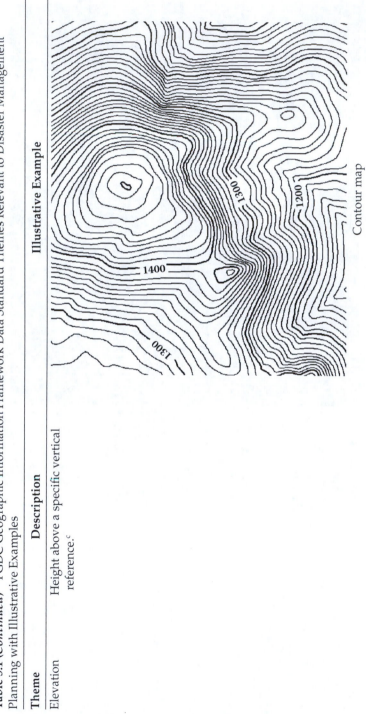

Contour map

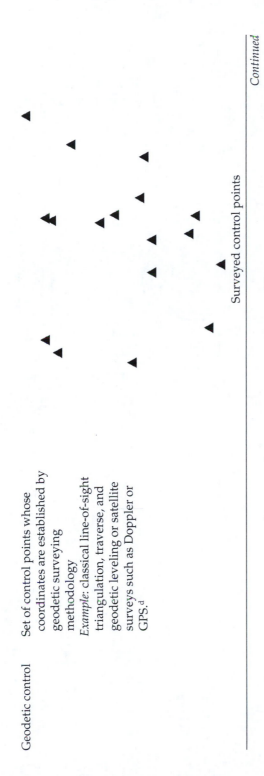

Geodetic control

Set of control points whose coordinates are established by geodetic surveying methodology
Example: classical line-of-sight triangulation, traverse, and geodetic leveling or satellite surveys such as Doppler or GPS.[d]

Surveyed control points

Continued

Table 5.1 (*Continued*) FGDC Geographic Information Framework Data Standard Themes Relevant to Disaster Management Planning with Illustrative Examples

Theme	Description	Illustrative Example
Government units and other geographic area boundaries	Boundary: set that represents the limit of an entity (may or may not follow a visible feature and may or may not be visibly marked).[e]	

Hydrography

Geographic locations, interconnectedness, and characteristics of features in the surface water system.[f]

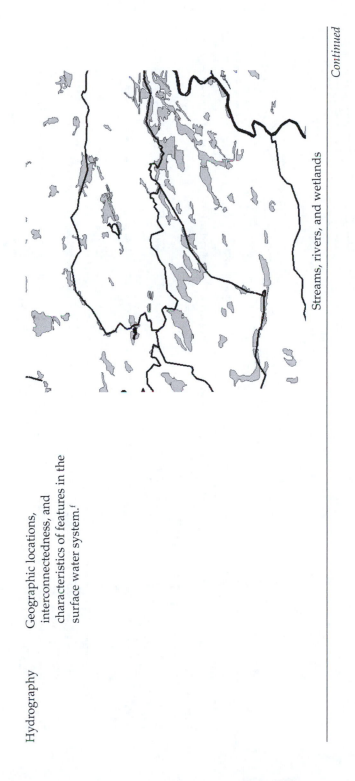

Streams, rivers, and wetlands

Continued

159

Table 5.1 (*Continued*) FGDC Geographic Information Framework Data Standard Themes Relevant to Disaster Management Planning with Illustrative Examples

Theme	Description	Illustrative Example
Transportation (including road, rail, inland waterways, public transportation)	Set of components that allow the movement of goods and people between locations.[g]	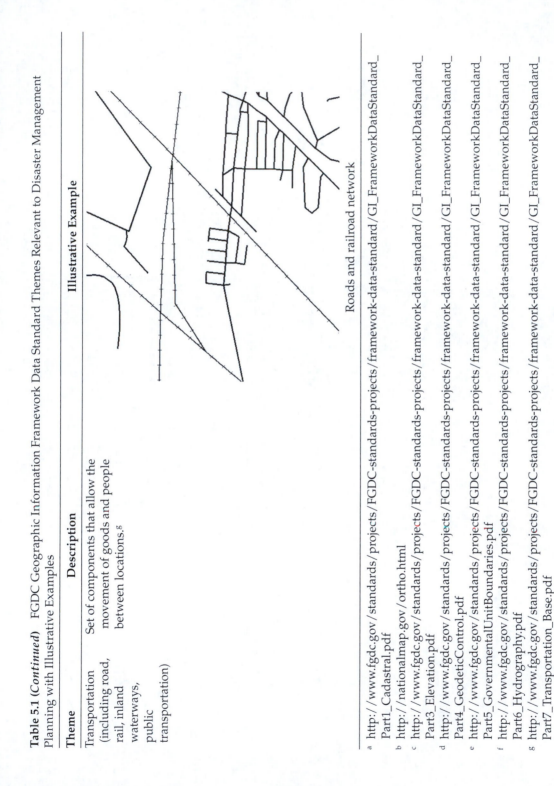 Roads and railroad network

a http://www.fgdc.gov/standards/projects/FGDC-standards-projects/framework-data-standard/GI_FrameworkDataStandard_Part1_Cadastral.pdf

b http://nationalmap.gov/ortho.html

c http://www.fgdc.gov/standards/projects/FGDC-standards-projects/framework-data-standard/GI_FrameworkDataStandard_Part3_Elevation.pdf

d http://www.fgdc.gov/standards/projects/FGDC-standards-projects/framework-data-standard/GI_FrameworkDataStandard_Part4_GeodeticControl.pdf

e http://www.fgdc.gov/standards/projects/FGDC-standards-projects/framework-data-standard/GI_FrameworkDataStandard_Part5_GovernmentalUnitBoundaries.pdf

f http://www.fgdc.gov/standards/projects/FGDC-standards-projects/framework-data-standard/GI_FrameworkDataStandard_Part6_Hydrography.pdf

g http://www.fgdc.gov/standards/projects/FGDC-standards-projects/framework-data-standard/GI_FrameworkDataStandard_Part7_Transportation_Base.pdf the term *data model* is used to describe how real-world entities are modeled in terms of database representations and relationships between entities and conceptual hierarchies. As an example of a conceptual hierarchy, transportation will have a subtype of roads, which in turn will have a subtype of local roads. Discussion of GIS data models in this context is beyond the scope of this book. However, resources are provided at the end of this chapter for you to follow up on this topic as you progress in your GIS and database modeling skills.

Department of Homeland Security Geospatial Data Model

The Department of Homeland Security Geospatial Data Model (GDM) is "a comprehensive framework for organizing features of interest to the homeland security community. The essential purpose of the GDM is to provide a means for sharing of geospatial information sharing between organizations and agencies whose primary responsibility it is to plan for, and respond to natural disasters and hostile events" (Federal Geographic Data Committee, 2009).

In Figure 5.1, a graphical representation of the GDM, make note of how the various data themes expand into subthemes. On the left side of Figure 5.1, the transportation theme has been expanded to show trails and then trail subtypes. There are literally hundreds of different subthemes within this overall data model, which make it impossible to reproduce in full printed form such as this book. Thus, this data model is a thorough and detailed representation of the wide variety of data layers that can be used for disaster management and homeland security. GIS organizations often use a data model like this as a starting point for creating their own in GIS databases that are then modified per the specific organizational context.

Technology Planning and Preparation

Although data is at the core of GIS, it is equally important that planning and preparation activities are conducted around GIS technology itself. For example, some commercial GIS technology requires regularly updated licenses to continue working with the technology, and require management of general IT computer issues such as operating system upgrades, virus protection, and other activities to keep computing infrastructure running. If you are the "GIS person" that has to work with an IT support person, it is very important to develop a good relationship with the IT support person so that your GIS technology is in place and ready to go when a disaster occurs. Additionally, the time afforded during the planning phase is a good time to stay up to date on the newest trends in GIS technology as new features are constantly being added, revised, and modified as the technology grows. Thus, it is important to stay up to date on new technology, new datasets, and other aspects of GIS that are potentially relevant to disaster management activities. The topic of staying up to date on GIS for disaster management activities is addressed further in Chapter 9.

ORGANIZATIONAL PERSPECTIVES

Another aspect of planning and preparation for GIS and disaster management is developing organizations so that they are able to collaborate and share data, people, tools, and other resources with one another when disaster occurs. As discussed in Chapter 1, sharing

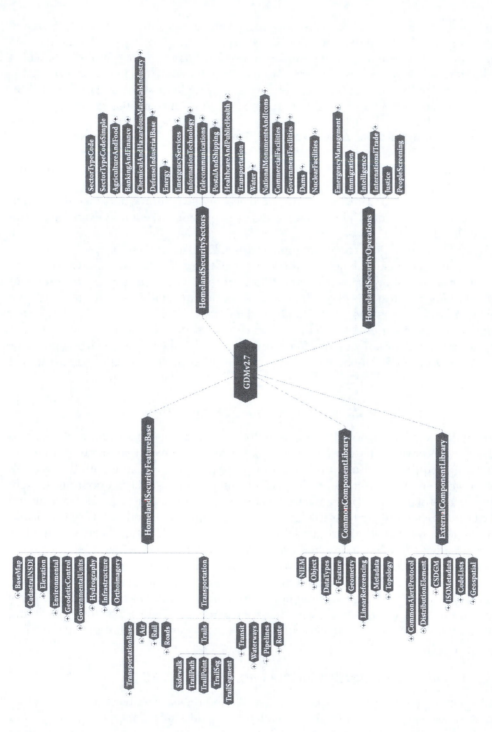

Figure 5.1 The Department of Homeland Security Geospatial Data Model. GDM Mind Map of GDM v2.7. (Image from the Federal Geographic Data Committee available at http://www.fgdc.gov/participation/working-groups-subcommittees/hswg/dhs-gdm/version-2-7 [accessed April 23, 2014].)

MEMORANDUM OF UNDERSTANDING

BETWEEN

The Town of Kennebunkport, Maine

AND

The Federal Emergency Management Agency

Working together on a Risk Mapping, Assessment, and Planning (Risk MAP) project, FEMA Region I and the Town of Kennebunkport, Maine, will identify, assess, communicate, plan for, and mitigate against flood risk. Updated flood risk information can enhance local and State hazard mitigation plans, and improve Kennebunkport's ability to raise awareness about flood risk, while making informed decisions that will improve its resilience to flooding.

This project MOU:
- o   Describes the long-term vision for the community or watershed as previously identified
- o   Describes the mapping, assessment, and planning information, products, and assistance that FEMA will provide over the course of the project
- o   Describes the milestones and sets the schedule for when those milestones will be achieved
- o   Summarizes local flooding concerns and provides a general idea of areas expected to have increased flood hazards
- o   Describes roles and responsibilities of FEMA and the community throughout the project

Figure 5.2 Memorandum of understanding example. (http://www.kennebunkportme.gov/ Public_Documents/KennebunkportME_Code/FEMAPrelimMaps/Kennebunkport%20MOU.pdf).

of data and other resources across multiple organizations spanning multiple jurisdictions continues to be a major hindrance to the effective use of GIS during disasters. Organizations that utilize GIS for disaster management activities must take steps to make sure data-sharing agreements and memorandums of understanding (MOUs) are in place *before* disasters occur so that data and information can be easily shared to support collaborative disaster management activities. This is particularly important for organizations in jurisdictions that have limited or different levels of GIS capacity as per the discussions in Chapter 4. For example, a town may wish to utilize data resources of the US federal government provided by FEMA, and in turn, new data created by the town can then be sent back to FEMA for broader dissemination. Figure 5.2 shows an example of an MOU between FEMA and the town of Kennebunkport, Maine, centered on shared risk mapping activities.

USING GIS TO SUPPORT PLANNING AND PREPARATION ACTIVITIES

Spatial Perspectives on Broader Planning and Preparation Activities

Although the use of GIS as a tool for disaster planning and preparation will be discussed momentarily in the next section, it is important to consider spatial perspectives on broader planning and preparation activities. Not all disasters are the same. It is important to consider the "geography" of the area such as people, the places they live, attachments to their communities, and relationships and interactions at multiple scales. This is another good example of the idea of spatial thinking that was presented in Chapter 1. GIS tools are good at doing what computers do best—representing things in a very discreet manner such as binary 1s and 0s. However, this can sometimes be a limitation in that GIS and the way it represents reality can be very disconnected from reality itself. Thus, it is important to keep in mind that GIS should not be the final "word" on how disaster management plan should be developed. Rather, GIS should be used as a support device that supports spatial thinking of disaster management practitioners, private citizens, and anyone else needing to think of the spatial and geographic nature of populations for disaster planning.

Common GIS Tasks for Disaster Planning and Preparation Activities

The following sections discuss common GIS disaster management planning and preparation tasks. Specific GIS technologies are not used to illustrate the examples. Rather, the general concepts and ideas are presented. You are encouraged to learn how to implement these ideas with the specific GIS technology with which you work.

Evacuation Route Planning

One of the most common uses of GIS for disaster management planning is development of evacuation routes. Understanding how and where to evacuate people during a disaster is a fundamental activity during a wide range of disasters such as hurricanes, wildfires, snow storms, and other events that require people to move quickly to safe areas. Figure 5.3 is a framework for evacuation route planning.

At the top of Figure 5.3 are reference spatial datasets. These refer to the underlying, essential data such as those discussed previously in this chapter, that are used for developing evacuation route plans. The most fundamental of these datasets are the transportation network, which most commonly are roads. Within a road network dataset, attributes that must be available include

1. the road type, such as single lane, multilane; the road category, such as residential street, highway, exit ramp and other categories; and the overall traffic volume the road is rated for;
2. the road direction, such as one way or multidirectional roads; and
3. the speed for which the road is rated for travel.

Other reference spatial datasets include traffic sensor data if it is available. This might include real-time road volume and real-time accident reporting, both of which can feed into rapid decision making. Additionally, traffic sensor data may include historical traffic data such as patterns of use, accident locations, and other factors that can be used as planning scenario inputs. The final reference spatial datasets are the broader category of context -specific data that is used in specific planning scenarios. For example, data on the built environments, such as building locations and population characteristics of people that will be difficult to evacuate, such as elderly and children who may not have access to vehicles, along with pets.

Based on these and other reference spatial datasets, analytics are then run to determine specific evacuation routes. These analytics are a very broad category that can include

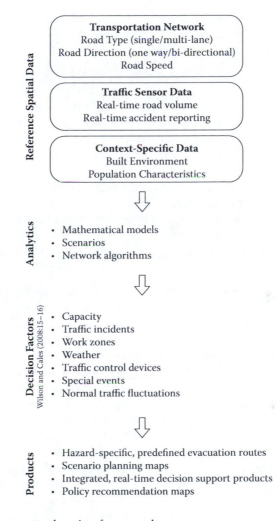

Figure 5.3 Evacuation route-planning framework.

mathematical models for solving problems such as congestion and bottlenecks that occur at traffic intersections, what-if scenarios for particular routing and evacuation situations such as a snow emergency or emergency evacuation during an event of mass gathering such as a sporting event, and network algorithms that are available in most commercial and open-source GIS tools with stronger analytic capabilities. These network algorithms typically perform functions such as defining routes based on least travel time, least travel distance, and other cost factors. Of particular importance for evacuation route planning are the ability of network algorithms to account for obstacles in barriers that may occur in travel routes. This is particularly important in time-sensitive situations where factors cannot be accounted for ahead of time, such as road blockages due to traffic bottlenecks or built environment impacts such as damaged streets. Chapter 7 discusses examples of network algorithms in the context of disaster recovery.

The reference spatial datasets, and the analytics operated on them, are then used as decision factor inputs for developing evacuation plans. Decision factors, based on Wilson and Cales (2008, 14–15), include

- capacity of the road in terms of the volume of traffic that can be handled in a given section of road and based on factors such as road with a shoulder;
- traffic incidents such as crashes and breakdowns and debris in travel lanes;
- work zones that create physical modification to the transportation network such as closing of lanes, shifting of lanes, or even temporary closures;
- weather that can impact driver behavior such as slower driving during rain or snow;
- traffic control devices such as railroad crossings and traffic signals that can factor into travel time variability and congestion;
- special events that are outside of typical traffic patterns such as sporting events or other large gatherings of people; and
- fluctuations in normal traffic such as workday versus weekend travel on road networks.

Based on these issues and other possible decision factors, final evacuation route planning products can be developed. For example, hazards-specific predefined evacuation route maps that can be given to disaster management practitioners and the general public (Figure 5.4).

Additionally, evacuation route analysis with GIS can feed into what-if scenario planning maps such as those used during planning for major events such as sporting events where alternative emergency evacuation planning may be different than normal circumstances. Decisions made can also factor into integrated real-time decision support products of which GIS is ideally suited as a common platform for incorporating data inputs from multiple sources (Wilson and Cales, 2008). This point is particularly relevant to rapidly unfolding situations where real-time decision support is essential to facilitate disaster response activities. For example, large-scale disasters such as Hurricane Sandy in 2012 required constant monitoring and updating of road closures and openings to make sure that people and relief supplies were moving in an expedited manner to support the response (Mainline Media News, 2013). Finally, policy recommendation maps can be developed to inform long-term planning on the designing and capacity of transportation networks to handle emergency evacuation scenarios. For example, and similar to the idea of a hazard-specific predefined evacuation route, these policies might include the placing of signs next to road signs to inform the public as to where to evacuate during an emergency (Figure 5.5).

Evacuation Zone Planning

Closely related to the evacuation route planning, evacuation zone planning is the idea of defining areas that are (1) to be evacuated during a disaster or (2) the areas to evacuate *to* during a disaster. For example, coastal areas that are prone to events such as hurricanes and tsunamis and subsequent storm surges may have different evacuation zone categories. These categories would be defined based on predicted storm surge levels that may impact the area. Areas directly next to the coast and at the lowest elevation would be

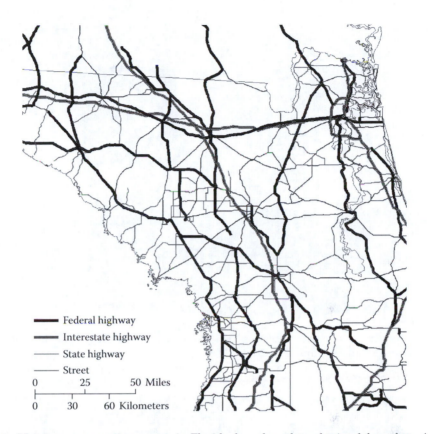

Figure 5.4 Hurricane evacuation routes in Florida, based on data obtained from http://gis.fema.gov/DataFeeds.html. (Map by Brian Tomaszewski.)

evacuated first, with additional evacuation zones being defined by moving farther away from the shore and based on elevation changes and storm surge heights. The zones can then be marked with signs to warn people of potential hazards (Figure 5.6).

Defining the evacuation zones that people would be evacuated to during an emergency can be based on a wide variety of factors such as shelter locations, elevation, access to medical facilities, connection to transportation networks, and any other context-specific factors.

Figure 5.7 is a simple, yet realistic GIS analysis site selection problem that might be conducted to find flood evacuation zones. The idea is to get you thinking of how specific GIS tools and datasets discussed in Chapter 3 can be utilized for conducting such an analysis. This should just be considered a starting point from which you can add your own, context-specific elements to investigate other more complex methods and datasets for modeling these scenario types.

Reference datasets to use include a digital elevation model, or DEM, to determine land heights that might potentially be impacted by a flood, locations of buildings, and transportation networks. The flood itself can be simulated using a buffer, the distance of which

Figure 5.5 Hurricane evacuation route sign. (FEMA photo, Jocelyn Augustino; English Wikipedia published under Creative Commons Attribution 3.0.)

Figure 5.6 Tsunami hazard zone site. (Wikipedia Commons. Unmodified photo taken by user Mimigu at English Wikipedia published under Creative Commons Attribution 3.0.)

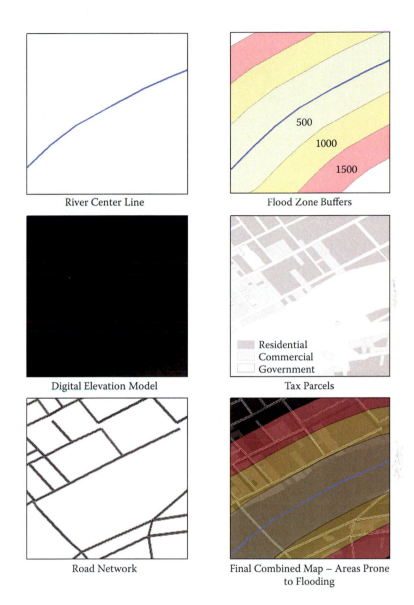

River Center Line

Flood Zone Buffers

Digital Elevation Model

Tax Parcels

Residential
Commercial
Government

Road Network

Final Combined Map – Areas Prone
to Flooding

Figure 5.7 A hypothetical yet realistic flood evacuation zone analysis. In this example, the individual squares represent specific data layers used to determine flood evacuation zones. Starting on the top left is the river center line, followed to the right by a multiring buffer that represents potential flood extent based on factors such as rain level. In the second row, starting on the left, is the digital elevation model that can be used for determining land height. In the second row on the right are tax parcels that can be used to determine building type and population characteristics. In the bottom row left is a road network that includes an important bridge that would be critical to an evacuation. The final map on the bottom right shows all the previous layers combined to support analysis of prioritized evacuation areas in the event of a flood.

represents the flood level based on factors such as rainfall. The flood buffer can then be overlaid on top of the DEM, buildings, and transportation network to determine which geographic features would be impacted by the flood to define the specific evacuation zones.

Scenario Modeling to Answer What-If Questions

Another very common use of GIS for disaster management and planning is to answer what-if questions. In a sense, the previous discussions on evacuation route and zone planning are examples of scenario modeling. However, the wide variety of potential disaster scenarios make GIS particularly useful for examining potential disaster scenarios that can be planned for in specific contexts.

As discussed in Chapter 3, excellent examples of GIS tools for scenario modeling are the Environmental Protection Agency's ALOHA plume modeling, FEMA's HAZUS tool, and the DHS SUMMIT environment. In fact, the HAZUS tool has the ability to scale from being a scenario modeling and planning tool to being a decision support tool when an actual event occurs, as seen in Figure 5.8, developed during the 2008 floods in Iowa.

Besides use of GIS tools for specific modeling, it is also the process of developing disaster scenarios. Disaster scenarios can be thought of like "stories" used to think

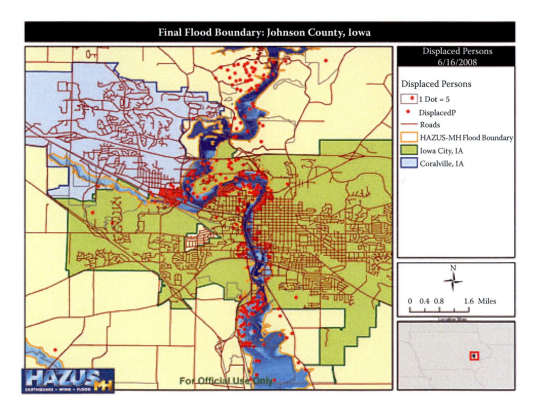

Figure 5.8 An example of a flood situation map developed using FEMA's HAZUS tool. (From FEMA image.)

through what would happen in the event that the scenario becomes reality. Developing your own scenarios for potential disaster situations is an excellent way to think spatially about disaster management activities and how GIS can support those activities. Scenarios in general are a very common technique used for training exercises that are often referred to as *tabletop exercises*. This is the idea of using a simulation to assess and test the ability of a group to respond to the given scenario. In the United States, FEMA has published a wide range emergency planning exercises that can be utilized for tabletop exercises by private sector organizations such as cyber attacks, earthquakes, chemical releases, and other situations. Many of these scenarios are based on the broader Department of Homeland Security (DHS) scenario planning document designed to provide credible natural disaster and terrorist attack planning scenarios for building disaster capacity and overall national preparedness (US Department of Homeland Security, 2006).

The following text, taken from the DHS planning document, describes the outline used in specific scenarios (US Department of Homeland Security, 2006, iv):

- Scenario Overview
  - General Description
  - Detailed Attack Scenario (or Detailed Scenario when a Universal Advisory (UA) is not present)
- Planning Considerations
  - Geographical Considerations/Description
  - Timeline/Event Dynamics
  - Meteorological Conditions (where applicable)
  - Assumptions
  - Mission Areas Activated
- Implications
  - Secondary Hazards/Events
  - Fatalities/Injuries
  - Property Damage
  - Service Disruption
  - Economic Impact
- Long-Term Health Issues

Note that in the scenario outline, spatial elements are inherent in the scenarios, such as the geographical dimensions and implications such as property damage, service disruption, and economic impact. Figures 5.9 and 5.10 are image excerpts from *Scenario 1: Nuclear Detonation—10-Kiloton Improvised Nuclear Device* where a nuclear device is detonated over a city and are presented to illustrate how maps are used for scenario planning and tabletop exercises.

Public Outreach and Citizen Participation

Another important use of GIS and GIS-created map products is public outreach and citizen participation in disaster planning and preparation activities. As you saw earlier in this chapter, besides using GIS as an analytic tool to develop evacuation routes, communication of those routes to the public in map-based formats is equally important. Another classic

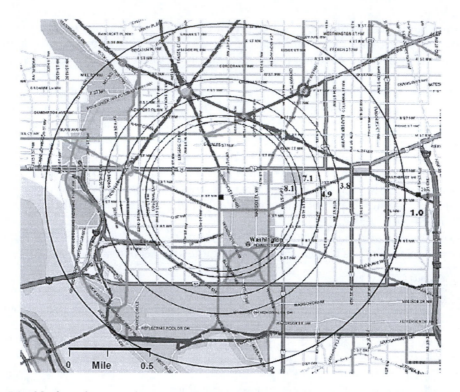

Figure 5.9 Nuclear disaster planning map. (From US Department of Homeland Security. 2006. National Planning Scenarios, 1–14.)

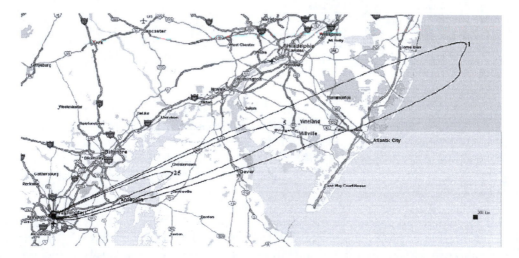

Figure 5.10 Nuclear disaster planning map. (From US Department of Homeland Security. 2006. National Planning Scenarios, 1–18.)

example of using maps in disaster planning activities is communicating the location of emergency shelters and their characteristics such as

- the number of people the shelter can accommodate,
- whether the shelter allowed pets,
- whether the shelter inside a flood zone,
- whether the shelter is accessible to disabled people, and
- other important criteria that need to be known ahead of time if people are to be evacuated to the shelter.

Many of these shelter characteristics can be communicated using qualitative map symbol strategies as discussed in Chapter 2. Furthermore, showing quantitative information is equally important and techniques such as graduated symbols, also discussed in Chapter 2, can be used to show overall shelter capacity and how many people are currently in a shelter. Figure 5.11 is an example of a shelter location map created from a FEMA shelter GIS data layer access through a web service. The number of different categories in which these shelters can be displayed is quite varied and Figure 5.11 shows basic shelter characteristics.

Online mapping tools also provide an excellent opportunity for communicating disaster planning information for public outreach planning. For example, tools like Google Maps Engine (discussed in Chapter 3) make it very easy to map basic disaster planning data such as shelter locations, evacuation routes, and other pertinent information online and easily accessible. However, caution should also be exercised because citizens may not have access to computers, the Internet, or even the ability to use and understand disaster planning and preparation information in map-based formats. Thus, online mapping tools that communicate disaster planning and preparation information should be considered one of several communication mediums you might use for communicating planning information. Low-tech, paper-based maps are still an equally viable way for public outreach communication as this medium is accessible to a very wide range of people (see examples in Chapter 7).

Mapping and GIS are also very powerful tools for incorporating citizens into the disaster planning and preparation process itself. For example, and as discussed previously in this book, the ideas of *crowdsourcing* can be utilized to allow citizens to collect data about their neighborhoods and communities that can be relevant when a disaster occurs. Along the lines of crowdsourcing and an excellent example involving citizens in spatial aspects of the planning process, is the Map Your Neighborhood (or MYN) project that was developed by the Washington State department of emergency management and received a FEMA Individual and Community Preparedness award in 2011. The MYN emphasizes the building of citizens' ability to operate during a disaster response, making connections with neighbors in their communities, and skill building such as understanding which supplies and resources are available during a disaster (Washington Military Department, Emergency Management Division, 2014). In terms of mapping, the program emphasizes the mapping of propane and natural gas locations. However, the ideas can also be extended to mapping any other critical or potentially hazardous source that should be identified during any disaster response.

In the GIS research world, there is also the idea of public participation GIS (PPGIS), which is a series of techniques for utilizing GIS as a means of involving the public in

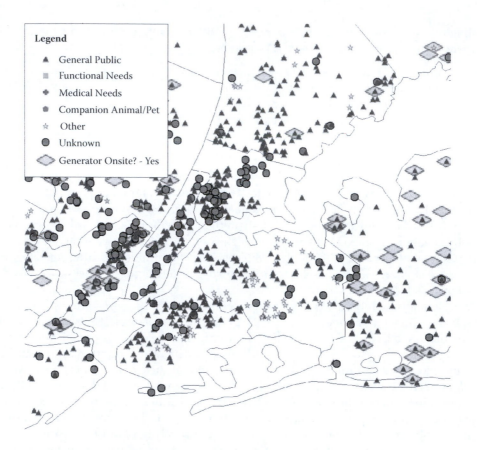

Figure 5.11 Map of shelter locations based on FEMA GIS data available through a web service. In addition to being a general example of showing shelter locations and their characteristics, it is also important to point out that this is a good example of designing a map that can be displayed in both color and black and white. For example, although all the points all have different colors associated with them, they also can communicate qualitative distinction through shelter shapes. By not relying solely on color, this map can be printed in black and white, and the meaning is not lost because of using the shape visual variable. (Map created by Brian Tomaszewski with data obtained from http://gis.fema.gov/REST/services/NSS/OpenShelters/MapServer.)

discussions related to planning and decision making, most often in urban planning contexts such as deciding how to redevelop a neighborhood based on input from multiple stakeholders and multiple decision criteria (Sieber, 2006; Jankowski et al., 2006).

Finally, in addition to basic disaster preparedness communications to citizens, is the idea of preparing citizens to think spatially and understand how to use maps and other spatial navigation devices during a disaster situation. As discussed previously in this book, the continued increase of devices such as smartphones with built-in GPS capability is creating a societal effect where many citizen are less capable of spatial navigation and reasoning without assistance from a GPS device. It is important that people still understand how to use paper maps as these are what will be available in the event that there is no power or ability to use GPS. In fact, in the United States, the ready.gov website outlines items for disaster supply kits and recommend including local maps, along with other basic essentials such as flashlights, batteries, drinking water, and food.

The final chapter section provides GIS and disaster management planning from a United Nations perspective to give you a sense that the disaster planning and preparedness challenges faced at all levels in the United States also exist at the international level.

GIS AND DISASTER MANAGEMENT PLANNING: A UNITED NATIONS PERSPECTIVE

Interview with Lóránt Czárán*

Lóránt Czárán (Figure 5.12) works with the United Nations Office for Outer Space Affairs (UN-OOSA) with a focus on space technology and its applications for disaster management, environmental monitoring, and natural resource management. As of 2014, however, he is currently on a temporary assignment with the United Nations Cartographic Section[†] in New York where he is focusing on the boundary demarcation between Cameroon and Nigeria. His responsibilities on this project center on producing final boundary map products for the governments of Cameroon and Nigeria as part of a boundary agreement. His educational background includes geography and Russian language studies at Cluj University in Romania. He began working with the UN in 1996 and he since developed significant expertise in GIS and remote sensing through previous positions with the UN Cartographic Section, United Nations Environment Programme (UNEP), United Nations Office for the Coordination for Humanitarian Affairs (UN-OCHA), and UN-OOSA. He is also very active in promoting the use of GIS within the UN system through his activities with the United Nations Geographic Information Working Group (UNGIWG).[‡]

Figure 5.12 Lóránt Czárán. (Photograph by Brian Tomaszewski.)

---

* The views expressed herein are those of the author(s) and do not necessarily reflect the views of the United Nations.
† United Nations, http://www.un.org/Depts/Cartographic/english/htmain.htm.
‡ United Nations Geographic Information Working Group (UNGIWG), http://www.ungiwg.org/.

The following is the first of a two-part interview conducted for this book with Mr. Czárán in January 2014. In this portion of the interview, he answers questions about his specific work with the UN Cartographic Section and the role of GIS in wider UN disaster planning activities. The second half of this interview is presented in Chapter 9 where Mr. Czárán provides advice on how to secure a job in the GIS industry for disaster management, and the future of GIS for disaster management.

Describe what the UN Cartographic Section does.

The Cartographic Section does not necessarily have a disaster management support mandate. We have other specialized agencies and units within the UN system with that specific mandate, but in the same time, the UN Cartographic Section is maybe the most important geospatial support unit within the UN Secretariat as it has a specific mandate in providing geospatial information support to departments and offices of the UN Secretariat,* to UN peacekeeping missions, and to various other UN entities on request. In that respect, the Cartographic Section has probably the most specialized staff in delivering on this mandate and support; the Section right now is part of the Department for Field Support, supporting peacekeeping missions by providing data, mapping, satellite imagery acquisition, and analysis for the peacekeeping personnel so that they are more familiar with the specific areas of their operations. But, in the same time, the Section also has a mandate to provide assistance to other departments of the UN Secretariat and to provide UN map products clearance as well. Or more specifically, making sure that every map that is produced by any UN Secretariat office or even other UN agencies conforms to the UN standards such as naming conventions and other elements. The Cartographic Section also has the responsibility to support the UN Security Council with geospatial information during presentations, maps display during the Security Council consultations on various crises around the world. So, that's also one important aspect. Because of this wide-ranging mandate and being the geospatial authority within the UN Secretariat, the Section often sees itself providing support to disaster management–related situations as well.

For example, let's take 2010 in Haiti, a well-known disaster. When the big earthquake in Haiti happened, we had a peacekeeping operation on the ground that had the geospatial support unit and the Cartographic Section was responsible for the geospatial program that unit also belonged to. So, in that sense, the Cartographic Section here was the one that activated the International Charter (on space and major disasters) and other support mechanisms after the earthquake, continuing with a lot of the support provided in the mission, by the mission GIS unit to other UN agencies and others who were deploying to Haiti, all under the coordination of the Section here. So, for example, mechanisms like G-MOSAIC† in the European Union and other support mechanisms in addition to the International Charter were activated through the Cartographic Section as well. Then, of course, the imagery data provision was also made easier through the contribution of the staff here and through contacts with the US State

---

* United Nations, https://www.un.org/en/mainbodies/secretariat/

† G-Mosaic, http://www.gmes-gmosaic.eu/.

Department and other partners that could provide support in that context. Haiti is one good example, of course, because we already had a peacekeeping mission on the ground and it was a major disaster, but in many other situations when OCHA or others require certain support and the Cartographic Section could provide that support in any way in terms of data sharing or sharing imagery or any other products, that's always being attempted. So, in that sense, the Section has an important role to play potentially also in disaster management situations, obviously. As long as it pertains to its mandate or that of any office or department in the Secretariat or even within the larger UN family, the Cartographic Section is there to provide support as possible.

Does the Cartographic Section do similar things like MapAction, such as going into the field and setting up rapid mapping services, or it's all really done from headquarters locations and disseminated to the field?

Rapid mapping is the domain of other units within the UN family. The Cartographic Section also works closely with NGOs [nongovernmental organizations]. We collaborate with the crowdsourcing community, OpenStreetMap, or the Google Map Maker team, or, let's say, with any relevant communities when it comes to supporting any of these disaster situations. So, no, the Cartographic Section does not have the capacity to actually deploy people in the field when a disaster happens. It's more indirect support and in terms of reaching out to a network of other institutions outside the UN too, institutions that we are collaborating with such as the US State Department or the EU (European Union) Satellite Center* or Esri and other private sector partners. Very often, we work together with them or other resources within the EU context when we have good contacts and links to offer; reaching out to those partners to enlist their support is also something that's being done from here. So, no, definitely not going to the field, but preparing helpful products, helping to brief the senior management of the UN here at the headquarters about the situations on the ground and generating map products in that sense is more what the Section does—supporting with that other UN entities such as OCHA, WFP, UNHCR or others when or if they need it. As an example, what is happening as of 2014, with Syria and the joint mission of the OPCW [Organisation for the Prohibition of Chemical Weapons] and UN[†] to assist in removing chemical weapons. All these activities, as much as possible, are supported from the Section with provision of map products and especially customized maps, datasets, and other activities.

Is the term "situation awareness" a term the Cartographic Section uses a lot?

Definitely, because that's one of the crucial needs in the context of a peacekeeping mission. We have approximately 20 or so peacekeeping and peace-building missions out there in the field. Many of them have geospatial support units within the mission structures, which help. The military component and other components in the mission have access to good maps, updated information, geospatial information

---

* European Union Satellite Centre, http://www.eusc.europa.eu/.
† Organisation for the Prohibition of Chemical Weapons, http://www.opcw.org/about-opcw/un-opcw--relationship/.

that helps their movements on the ground. And in each context, of course, it's not only the provision of maps or data, but it's also how to exploit that data and how to exploit those products in a way to better support the decision making, to have that situational awareness on the ground for the troops and for the other components of the missions that operate in those difficult areas. That's clearly an aspect that the Section tries to cover from here. Many products and many projects over the past number of years have been aimed at that. For example, the acquisition and development of the so-called UN Earth, aimed to set up a decision support environment whereby Google Earth server* software was acquired and deployed in a number of major missions for developing their own internal visualization of geospatial data, to better support their situational awareness and decision making. That's one example, and we are continuously aiming at more advanced decision support systems, situational awareness improvements that could benefit the most important UN operations on the ground. Of course, we ideally should be collaborating with other organizations in the UN family and outside the system, with those who are trying to do the same, and that will be the next big step, if not challenge, in terms of getting the job done better.

What is your specific GIS mapping work in terms of disaster management as part of your duties with the UN-OOSA?

My permanent position that is in Vienna with the Office for Outer Space Affairs takes me to countries, especially on the African continent where we are organizing a number of technical advisory missions to identify the needs countries have in terms of space technologies and geospatial information when it comes to disaster management. Then based on those needs identified, we try to organize capacity building, training, workshops for the national authorities in charge of disaster management on the ground to get more familiar with geospatial technologies and remote sensing data and to learn how to apply that in their work. Of course, as often as needed, I also support disaster situations in the emergency response phase in those countries if they reach out to us.

How does your Cameroon-Nigeria boundary demarcation work fit with disaster management?

Well, what I'm doing here is developing the so-called final mapping of this boundary line, which is a thousand and almost eight hundred kilometers long boundary. Most of it runs through very difficult environments, such as mountain areas, that are very little explored, or inaccessible, with no roads, no settlements. It's a very challenging task to properly identify the border based on the international treaties and other legal mechanisms in place. This is a job that has been going on for the last seven or eight years and right now, we are in the final phase.

An important issue to mention on boundary demarcation is, in my view, that a lot of this geospatial data that we develop and that would be handed over to the two countries in the future, although it is mostly in a buffer along the boundary to be *delineated and mapped*, would be maybe useful for other applications as well because it is highly accurate, high resolution. A lot of resources were spent in purchasing and rectifying satellite imagery and developing derived

---

* Google, http://www.google.com/enterprise/earthmaps/earth_technical.html.

geospatial data. Once it is all handed over to the two governments, of course, it could be well used in disaster management–related situations as well or to just enrich the national spatial data infrastructure of those countries. We are also looking at obtaining very-high-resolution imagery for certain areas where the border crosses villages and where you clearly need a higher level of detail. We are also reaching out to traditional partners such as the US State Department, National Geospatial Intelligence Agency (NGA) and other actors that could support us in getting access to better data such as the 30-meter SRTM 2 dataset for the boundary area. Once available, those datasets could all be beneficial for the two countries as well if we manage to obtain their release and hand-over as well. So, in that sense, indirectly everything we do here might serve them well as reference data in some future border or cross-border collaboration or disaster management situations as well.

A lot of times there is talk about the dangers posed by Mount Cameroon in the south of Cameroon and relatively close to the border with Nigeria also, in case that volcano would erupt, or other situations like around Lake Nyos, which is a very dangerous lake that had the gas trapped under it and has a history of smaller-scale disaster.* Given that Lake Nyos is also very close to the border area, both countries have an interest in having that area well mapped and monitored. Thus, the fact that we developed this geospatial data or that we work on such data, even if it might be strictly for a boundary demarcation now, such resulting data might in the future perhaps serve in other contexts as well. So, in a sense, planning—getting things in place, being prepared if something happens. Whatever you develop, whatever you produce might be used in multiple contexts. Our results on this final mapping project do not have to be looked at only as, say, a set of nicely printed boundary maps, but as well, we also have geospatial data that could be part of a developing *spatial* data infrastructure in this or that country. That might be beneficial later for other reasons. And the continuous drive to obtain better, high-resolution data, digital elevation image data also can help in other contexts, even though now our priority is that specific boundary mapping.

Could you tell me a little bit about UNGIWG?

UNGIWG is in fact the UN Geographic Information Working Group. It is an informal but very important internal collaboration mechanism between all the UN family organizations when it comes to geospatial information management, mechanism that was set up as a working group in 2000. Since then, it has met every year. Today, UNGIWG has about 35 UN entities, organizations part of it and about 600 geospatial experts across these 35 organizations that are part of a mailing list that we use to communicate and to encourage collaboration between the different entities when it comes to use and development of geospatial data. I mention it because discussing issues like common operational datasets, the word common already means that you need to make sure that everybody's on the same page. It's not enough that one agency, even though they have a mandate, comes up with the data requirements or a list, everybody else needs to agree too,

---

* How Stuff Works, http://science.howstuffworks.com/environmental/earth/geophysics/lake-nyos.htm.

to contribute and make sure that it's a feasible and valid list of requirements. That's what makes it common and that's the spirit of collaboration we need.

How does the UN plan for disasters in terms of GIS data support and services and such?

Well, of course, the UN is one big family. We have different organizations, different structures, and different mandates. So, it's difficult to generalize. But, I can speak from my experience of working both in peacekeeping and OCHA when I was with the ReliefWeb,* then in UN-OOSA where we are tasked by the UN General Assembly with supporting governments for being better prepared for disaster response. So, looking back at all those different positions and responsibilities, of course, I have a fuller picture of this planning that we are doing. It depends, again, on the mandates. Some organizations are strictly geared towards supporting in the emergency response phase. And that's a different set of planning and preparation that you do in that context than it is to work with training or work with early-warning type of activities. So, again, it's hard to give one specific response, but in essence I will say that any of us or any of the UN entities or UN departments or agencies that we are talking about would primarily look at how to develop a good, useful baseline data and a set of connections, a set of partnerships that would help by being activated when something requires it. So, in that sense it's interesting to mention the Common Operational Datasets (CODs) effort that OCHA is making.† These are standard datasets needed in any disaster situation, and they are being made readily available for all countries. For example, transportation and infrastructure networks, good elevation data, population data, statistical data, all the administrative boundaries which are very clearly and accurately maintained. So, data that is very useful and is needed any time something happens and you don't want to scramble when something happens. This has taken a long time to realize, and we are still in the process of developing it, developing such geospatial data, but this is definitely one key planning aspect. The goal is to have a common set of data that anybody has access to, can use, it's sharable and can be easily exploited when the situation requires it.

Colleagues are working constantly on making these data available in standard formats. They are compiled and put online on certain servers so that anybody can download them. They might be in proprietary formats, but in any case as standardized as possible and as open and freely available as possible. That's the *idea*. They are also available to the public. Primarily, of course, it's important for the UN agencies to be on the same page, but the idea is also to have data available publicly because then a lot of NGOs and other partners, external partners that we work with would have access to the same data. And then, they could and would all work off the same data. So, this effort is important, humanitarian common operational datasets is one step in the right direction for sure. I would think that other agencies, they should contribute more to that effort and other similar initiatives, contribute their data holdings and have these

---

* ReliefWeb, http://reliefweb.int/.
† Humanitarian Response, https://cod.humanitarianresponse.info/.

discussions maybe in a larger context, such as the UNGIWG context to see how these common datasets can be improved, how their accuracy, resolution or scale could be improved. Our more active participation in specialized and dedicated forums such as the Group on Earth Observations [GEO] Data Sharing Working Group [DSWG] would also help.

Furthermore in terms of disaster planning, basically, it's the importance of available baseline, good data over any area where we would potentially have an increased risk of disasters, especially because some areas, some countries, some regions of the world, are not as well mapped as others, and data for those areas might not be as readily available. We might have situations where a country's internal administrative boundaries have not been updated for years because of a given situation. Take Somalia, for example. You don't even have a concrete governmental source to ask for such geospatial data or information because of the prolonged war and situation internally. So, in these situations, it's more diffi-cult to reach that ideal standard, accurate dataset target. But as much as possible, efforts should be made in that sense. So, that's key, of course.

Making sure that we have these partnerships in place is another effort. From our perspective, of course, looking at satellite imagery and other data providers, be it commercial entities or governmental agencies, space agencies, we have to have standby agreements with them, which would enable us to quickly request tasking of satellites or activation of certain mechanisms so that in any disaster situation, instead of waiting for days or weeks to get access to some new imag-ery over an area of interest, we can have these arrangements in place and would allow us to have access within a day or two to that data so that we can imme-diately make assessments, estimates, and help support our colleagues on the ground and those who work on the ground to respond, for example. So, that's another aspect that motivates many of us in the UN in working towards build-ing these partnerships. Building partnerships is also important because some-times you don't have the resources internally to act or to respond quickly in any situation. In that sense, we have the need to work with our external partners and to sign up new institutions that could support us. For example, the UN-SPIDER program [discussed in Chapter 4] has been enlisting voluntary Regional Support Officers (RSOs) in its network, expert organizations in this domain. We have been working closely with NASA and other space agencies worldwide to be able to call upon their support as soon as something happens in terms of tasking a satellite sensors quickly, informally even before some more formal mechanisms kick in. That's important as well because we could sometimes get a day or two faster access to whatever imagery is collected over a disaster area. So, that's another aspect of planning.

Having some internal coordination mechanisms is important as well. There is, for example, an interagency standing committee (IASC) that involves UN agen-cies, NGO representatives, and these representatives are always involved in dis-cussing the way the international assistance system would respond to any disaster situation—either man-made or natural disasters. Discussions within such working groups are also key to the planning of how to respond, how to react to situations.

That interagency standing committee has, I think, a subgroup, a working group that relates to information management, and in that context geospatial information is also discussed and issues such as the common operational datasets are brought up on a regular basis. So, those mechanisms are also important in terms of planning and preparing. There are also others, mechanisms such as the Central Emergency Response Fund [CERF]* that various countries put together so the UN can faster react to any major disaster situations. I would love to see provisions for that fund to concretely support acquisition of satellite or geospatial data, just the same way as today the same fund supports acquisition of food or shelter—or other materials. Have provisions for rapid new data acquisition, data management as well in the context of these funding mechanisms that are designed to react quickly when a disaster happens, because that's where we are still lagging way behind. For some reason, investment in data is never a priority. Sometimes senior management expects it to just appear for free. It might, but maybe in a week or maybe in two. And by that time it's too late for a number of reasons. And I'm convinced that if such funding mechanisms could be used for the quick data procurement just as they are used today for other elements, we might make much better progress in terms of using geospatial data for supporting the disaster management phases and everything in that context.

I also think closer collaboration between UN entities and partners is important and could still improve. UNGIWG is an informal mechanism and sometimes it doesn't work because it's on a voluntary basis and everybody is more interested in their own priorities. Evolving some of these collaboration, coordination mechanisms into more formal and having senior management more involved and ensuring that any available data is quickly and smoothly exchanged, that geospatial data is fully shared, with everybody knowing who's doing what in such a situation, would all be very beneficial. These actions are all key in terms of better planning, better response, avoiding any duplication of efforts and avoiding situations like what happened in Haiti when in the context of two or three weeks, over 1,500 map products were developed by various organizations, but then maybe not even a tenth of those were literally used by officials in the context of the emergency response. So, these are situations where commendable efforts are being made, but there's still room for improvement in terms of planning for any such situations.

SUMMARY

In this chapter, you learned about the relationship between GIS and disaster planning and preparedness. You first learned about perhaps the most important aspects of planning and preparation of GIS for disaster management—the planning and preparation of GIS datasets and GIS technology itself. More specifically, you were shown examples of GIS data frameworks that can be used for designing databases of relevant map layers and other

---

* UN CERF, http://www.unocha.org/cerf/.

spatial data assets for use in other disaster management phases. Next, you were given some perspectives on how organizations can prepare in terms of developing memorandums of understanding and cooperation agreements that can be enacted when a disaster occurs. These are critical to have in place before disaster strikes so that collaboration and coordination can occur to facilitate the best and most effective response possible or to support activities such as disaster recovery or mitigation. You were then given some ideas on how GIS can specifically be used as a support tool for planning and preparation activities. This discussion started with perhaps the most common disaster planning and preparation uses of GIS—development of evacuation routes and zones. You were also given some ideas on how to go about developing scenarios and simulations that can be used to answer what-if questions for planning purposes. As discussed, scenarios and simulations are very important training tools and activities for developing disaster capacity and readiness. You were then given some ideas on how GIS can serve as a public outreach and citizen participation tool for disaster planning and preparation activities. On this topic, you saw examples of developing maps to communicate to the public about shelter locations and their characteristics as per the Chapter 2 cartographic design ideas. You were also given a perspective on using GIS as a public participation tool, for example, using maps and GIS to gather input from citizens about their neighborhood, so that when an emergency occurs, people are aware of where potential hazards are located in their neighborhoods (such as propane tanks) and they have made connections with their neighbors for mutual support. The chapter concluded with an extensive interview of GIS in disaster management planning and preparation from the international perspective of the United Nations. As you saw in the interview, many of the issues related to GIS in disaster planning and preparation exist at the international scale as much as they do at the US local, county, state, or federal level. Important takeaway points from the interview also echo topics that were discussed in the chapter such as the critical importance of having common datasets available and accessible and the importance of building relationships and networks that can be utilized during a disaster.

The next chapter follows the sequence of discussing the relationship between GIS and specific disaster management cycles with a particular focus on GIS and disaster response.

DISCUSSION QUESTIONS AND ACTIVITIES

1. Do you think the data themes represented in Table 5.1 are sufficient as reference layers for all hazard and disaster types, or do you think other data themes should be included?
2. Using the Internet, do a search on GIS training for disaster management. Is what you're finding comprehensive; if not, what do you think is missing?
3. Look through the web pages of your local government and see if you can find any examples of memorandums of understanding, cooperation agreements or anything else that can demonstrate how specifically your local government is planning for disaster activities and GIS. If you cannot find anything in your local government, look to your state government or even federal government.
4. Using the ideas presented in the evacuation zone planning section of this chapter, come up with a hypothetical disaster planning scenario that can be supported

with GIS. Try to think of things other than floods, because this was the example used in this chapter. Think about the local context where you live and the types of natural or other types and hazards that are relevant to that context.

5. What are some specific ways GIS could be used directly by citizens themselves as a disaster management planning tool?

RESOURCES NOTES

For more on GIS data modeling, see Arctur, David, and Michael Zeiler. 2004. *Designing Geodatabases: Case Studies in GIS Data Modeling*. Redlands, CA: Esri Press.

For more information on the Department of Homeland Security Geospatial Data Model Version 2.7, see http://www.fgdc.gov/participation/working-groups-subcommittees/hswg/dhs-gdm/version-2-7.

An Esri discussion of GIS for planning and preparation can be found at http://www.esri.com/industries/public-safety/emergency-disaster-management/gis-used.

The National Transportation Atlas Database (NTAD) is an excellent source of transportation datasets available from the United States Department of Transportation that can be used for evacuation planning. This database is located at http://www.rita.dot.gov/bts/sites/rita.dot.gov.bts/files/publications/national_transportation_atlas_database/index.html.

Esri's Network Analyst is a commonly used commercial GIS tool for network and graph analysis planning: http://www.esri.com/software/arcgis/extensions/networkanalyst.

QGIS also has a network analysis library and can be found at http://www.qgis.org/en/docs/pyqgis_developer_cookbook/network_analysis.html.

GIS data models in terms of modeling real-world entities in a database:
Storm Surge: http://www.nauticalcharts.noaa.gov/csdl/stormsurge.html

Hurricane Impact Analysis from FEMA: http://www.arcgis.com/home/item.html?id=307dd522499d4a44a33d7296a5da5ea0

Hawaii tsunami zone: http://www.scd.hawaii.gov/

For more discussion of FEMA scenario activities, see http://www.fema.gov/emergency-planning-exercises.

For more information on the Map Your Neighborhood (MYN) project, see http://www.emd.wa.gov/myn/index.shtml.

For more information about citizen preparedness kits, see http://www.ready.gov/basic-disaster-supplies-kit.

REFERENCES

Federal Emergency Management Agency (FEMA). 2004. *Are You Ready? An In-depth Guide to Citizen Preparedness*. FEMA, http://www.fema.gov/pdf/areyouready/areyouready_full.pdf (accessed June 2, 2014). Federal Geographic Data Committee. 2008. "Geographic Information

Framework Data Standard," Federal Geographic Data Committee, http://www.fgdc.gov/ standards/projects/FGDC-standards-projects/framework-data-standard/framework-data-standard (accessed April 5, 2014).

Federal Geographic Data Committee. 2009. "Department of Homeland Security Geospatial Data Model," Federal Geographic Data Committee, http://www.fgdc.gov/participation/working-groups-subcommittees/hswg/dhs-gdm (accessed April 5, 2014).

Jankowski, Piotr, Steven Robischon, David Tuthill Timothy Nyerges, and Kevin Ramsey. 2006. "Design considerations and evaluation of a collaborative, spatio-temporal decision support system." *Transactions in GIS* 10 (3):335–354.

Mainline Media News. 2013. "Timeline: Superstorm Sandy cuts power, blocks roads in 2012," Mainline Media News, October 23, http://www.mainlinemedianews.com/articles/2013/10/23/test_do_not_publish_here/doc5267f59754eca045525939.txt (accessed February 12, 2014).

Phillips, Brenda D., David M. Neal, and Gary R. Webb. 2012. *Introduction to Emergency Management*. Boca Raton, FL: CRC Press.

Plumer, Brad. 2012. "Sandy shows the U.S. is unprepared for climate disasters," *Washington Post*, October 31, http://www.washingtonpost.com/blogs/wonkblog/wp/2012/10/31/why-the-united-states-is-so-unprepared-for-climate-disasters/ (accessed April 6, 2014).

Sieber, Renee. 2006. "Public participation geographic information systems: A literature review and framework." *Annals of the Association of American Geographers* 96 (3):491–507.

US Department of Homeland Security. 2006. National Planning Scenarios.

Washington Military Department, Emergency Management Division. n.d. "Why map your neighborhood?" Washington Military Department, Emergency Management Division, http://www.emd.wa.gov/myn/myn_why.shtml (accessed February 9, 2014).

Wilson, Robert D., and Brandon Cales. 2008. "Geographic information systems, evacuation planning and execution." *Communications of the IIMA* 8 (4).

6

Geographic Information Systems and Disaster Response

CHAPTER OBJECTIVES

Upon chapter completion, readers should be able to

1. understand the basics about disaster response policy in the United States,
2. discern various geographical aspects of situation awareness,
3. understand what spatial data deluge means and identify various Geographic Information Systems (GIS) techniques for handling spatial data deluge during a disaster response,
4. be familiar with concepts related to disaster response GIS product development,
5. understand how GIS can be used for disaster response damage assessment tasks,
6. be familiar with various GIS technology used for field data collection, and
7. identify opportunities for and barriers to incorporating the public and volunteers in disaster response mapping through crisis mapping.

INTRODUCTION

Disaster response is perhaps the most widely known disaster management cycle phase outside of professional disaster management practitioner communities. For example, when very large disasters or even catastrophes occur, images of destroyed buildings, burning streets, and displaced people are often what the news media shows. In fact, such representations actually have the effect of creating myths about what disaster response actually is and the kinds of tasks in which emergency managers, first responders, and other disaster management practitioners engage. Most disasters generally do not have high levels of chaos and the people affected generally display a high level of resilience and calm (Phillips, Neal, and Webb, 2012).

In this same regard, disaster response is also the most publically visible use of GIS and mapping in general. For example, natural hazards such as hurricanes are often portrayed in the news media using maps that show the estimated time of landfall and these maps serve as a type of citizen early warning (Figure 6.1).

Furthermore, maps and GIS are often used by the news media to portray how a disaster situation and its subsequent response are being handled by various disaster management practitioners. Finally, maps and GIS can conjure images of large command centers and emergency operation centers with large screens portraying maps that are the embodiment of situation awareness and coordination activities. In this chapter, these and other ideas are explored. Like the other chapters, the incredibly wide variety of potential

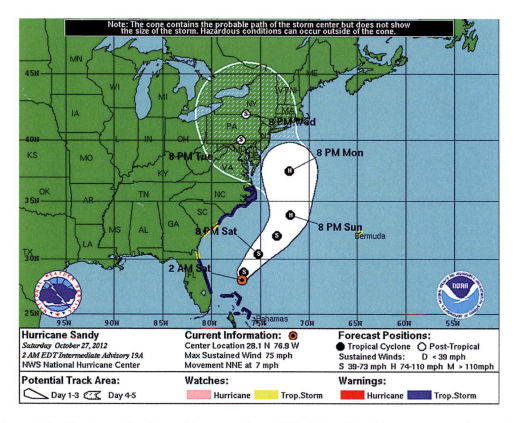

Figure 6.1 Hurricane Sandy tracking map from 2012. Maps like this are commonly used to portray disaster situations as they unfold and serve to inform and alert disaster responders and the public about oncoming disasters. In terms of map design ideas you learned about in Chapter 2, make note of several things in Figure 6.1. The first is the use of symbol orientations to convey uncertainty as to the potential track in days 4 and 5 over Virginia, Washington, DC, Pennsylvania, New York and some of the New England states. The second is the use of color to signify various types of watches and warnings (which of course will not be available if this map is printed in black and white). Finally, note how time references are added to indicate how the hurricane will progress. This map is a static image taken from an animated map of the storm trajectories posted on the NOAA website. (From NOAA, http://www.nhc.noaa.gov/archive/2012/graphics/al18/loop_5NLW.shtml.)

disaster situations make it difficult to predefine all types of GIS and disaster response activities. Thus, this chapter takes the approach used previously of presenting common GIS and disaster response ideas to give you perspectives on how these ideas might be applied your specific context.

The next section looks closer at disaster response policy in the United States with a particular focus on GIS as a follow-up to the policy and planning ideas you learned about in Chapter 4. Although from the perspective of the United States, the policy ideas are applicable to other contexts and can provide perspectives in terms of how GIS and disaster response activities are operationalized at the policy level.

Disaster Response Policy in the United States

In the United States at the federal level, disaster response is formulated into policy via the National Response Framework (NRF). In its own words, the NRF is "a guide to how the Nation responds to all types of disasters and emergencies. It is built on scalable, flexible, and adaptable concepts identified in the National Incident Management System [discussed in Chapter 4] to align key roles and responsibilities across the Nation. This Framework describes specific authorities and best practices for managing incidents that range from the serious but purely local to large-scale terrorist attacks or catastrophic natural disasters" (United States Department of Homeland Security, 2013, i). Updated in 2013, a particular emphasis was made on the "whole community" as an integral part of disaster response activities in terms of including individuals and families as important response activity components that center on "the capabilities necessary to save lives, protect property and the environment, and meet basic human needs after an incident has occurred" (Federal Emergency Management Agency [FEMA], 2013).

The NRF defines 14 core response capabilities, of which 11 are specific to response and 3 are common to all other FEMA disaster planning mission areas (prevention, planning, mitigation, and recovery). Furthermore, many of these core response capabilities are inherently spatial in nature, such as transportation, infrastructure, and situation assessment, thus making these and other spatially oriented examples relevant for GIS analysis and representation. The 14 core response capabilities are summarized in Table 6.1.

In terms of GIS and the fourteen core response capabilities outlined in Table 6.1, GIS is explicitly mentioned as an integrating technology as reflected in this quote: "The core capabilities in various mission areas may also be linked through shared assets and services. For example, the functionality provided by *Geographic Information Systems* can be applied across multiple response core capabilities, as well as core capabilities in the other four mission areas" (United States Department of Homeland Security, 2013, 25; italics added).

Another important aspect of the official role of GIS within official disaster response policy can be found inside some of the fifteen emergency support function (ESF) annexes of the NRF. ESFs "are mechanisms for grouping functions most frequently used to provide Federal support to States and Federal-to-Federal support, both for declared disasters and emergencies under the Stafford Act and for non-Stafford Act incidents" (Federal Emergency Management Agency (FEMA), 2008, ESF-i).

Table 6.1   NRF Core Response Capabilities

Capability	Tasks
1. Planning[a]	Conduct a systematic process engaging the whole community, as appropriate, in the development of executable strategic, operational, and community-based approaches to meet defined objectives.
2. Public Information and Warning[a]	Deliver coordinated, prompt, reliable, and actionable information to the whole community through the use of clear, consistent, accessible, and culturally and linguistically appropriate methods to effectively relay information regarding any threat or hazard and, as appropriate, the actions being taken and the assistance being made available.
3. Operational Coordination[a]	Establish and maintain a unified and coordinated operational structure and process that appropriately integrates all critical stakeholders and supports the execution of core capabilities.
4. Critical Transportation	Provide transportation (including infrastructure access and accessible transportation services) for response priority objectives, including the evacuation of people and animals, and the delivery of vital response personnel, equipment, and services to the affected areas.
5. Environmental Response/Health and Safety	Ensure the availability of guidance and resources to address all hazards, including hazardous materials, acts of terrorism, and natural disasters, in support of the responder operations and the affected communities.
6. Fatality Management Services	Provide fatality management services, including body recovery and victim identification, working with state and local authorities to provide temporary mortuary solutions, sharing information with Mass Care Services for the purpose of reunifying family members and caregivers with missing persons/remains, and providing counseling to the bereaved.
7. Infrastructure Systems[b]	Stabilize critical infrastructure functions, minimize health and safety threats, and efficiently restore and revitalize systems and services to support a viable, resilient community.
8. Mass Care Services	Provide life-sustaining services to the affected population with a focus on hydration, feeding, and sheltering of those with the most need, as well as support for reunifying families.
9. Mass Search and Rescue Operations	Deliver traditional and atypical search and rescue capabilities, including personnel, services, animals, and assets to survivors in need, with the goal of saving the greatest number of endangered lives in the shortest time possible.
10. On-scene Security and Protection	Ensure a safe and secure environment through law enforcement and related security and protection operations for people and communities located within affected areas and for all traditional and atypical response personnel engaged in lifesaving and life-sustaining operations.
11. Operational Communications	Ensure the capacity for timely communications in support of security, situational awareness, and operations by any and all means available between affected communities in the impact area and all response forces.
12. Public and Private Services and Resources	Provide essential public and private services and resources to the affected population and surrounding communities, to include emergency power to critical facilities, fuel support for emergency responders, and access to community staples (e.g., grocery stores, pharmacies, and banks) and fire and other first response services.

Table 6.1 (*Continued*)   NRF Core Response Capabilities

Capability	Tasks
13. Public Health and Medical Services	Provide lifesaving medical treatment via emergency medical services and related operations, and avoid additional disease and injury by providing targeted public health and medical support and products to all people in need within the affected area.
14. Situational Assessment	Provide all decision makers with decision-relevant information regarding the nature and extent of the hazard, any cascading effects, and the status of the response.

Source: United States Department of Homeland Security. 2013. *National Response Framework*, 2nd edition, FEMA, http://www.fema.gov/national-response-framework (accessed June 7, 2014), 20–24.
[a]   Cross-cutting with all mission areas.
[b]   Cross-cutting with recovery mission area.

GIS is specifically mentioned in Emergency Support Function 5, "Emergency Management":

- "Planning Section staff provide, manage, and organize geospatial data" (Federal Emergency Management Agency (FEMA), 2008, ESF#5-1).
- "Coordinates the use of remote sensing and reconnaissance operations, activation and deployment of assessment personnel or teams, and geospatial and Geographic Information System support needed for incident management" (Federal Emergency Management Agency (FEMA), 2008, ESF#5-6).

Emergency Support Function 9, "Search and Rescue Annex," also makes explicit mention of the Department of Defense National Geospatial-Intelligence Agency (NGA) as an organization that can use geospatial intelligence such as imagery and geographic analysis to support search and rescue operations.

Emergency Support Function 11, "Agriculture and Natural Resources Annex," also makes explicit mention of GIS and related activities. The first is the fact that the US Department of the Interior (DOI) is the lead for Natural and Cultural Resources and Historic Properties (NCH) Protection Policies and has responsibilities that include "up-to-date geospatial data related to impacted NCH resources, and develops and provides standard operating procedures for collecting NCH digital data, conducting GIS analyses, and disseminating geospatial products related to NCH resources, such as maps" (Federal Emergency Management Agency (FEMA), 2008, ESF#11-5). The second is that the Department of the Interior/US Geological Survey Animal and Plant Disease and Pest Response "assists in responding to a highly contagious/zoonotic disease, biohazard event, or other emergency involving wildlife by providing: wildlife emergency response teams; geospatial assessment and mapping tools; assistance in the identification of new emerging and resurging zoonotic diseases" (Federal Emergency Management Agency (FEMA), 2008, ESF#11-11). The third is that the Department of Agriculture/Food Safety and Inspection Service and Commercial Food Supply Safety and Security "[p] rovides Geographic Information Systems mapping capability for the meat, poultry, and egg product facilities it regulates to assist State, tribal, and local authorities to establish

191

food control zones to protect the public health" (Federal Emergency Management Agency (FEMA), 2008, ESF#11-11).

According to Emergency Support Function 13, "Public Safety and Security Annex," the National Aeronautics and Space Administration (NASA) "may utilize NASA assets and capabilities, such as geospatial modeling and decision support systems, aircraft with sensors, unmanned aerial vehicles, and a search and rescue team. These assets are designed to support a NASA event or NASA properties, but may be provided if requested for ESF #13 missions" (Federal Emergency Management Agency (FEMA), 2008, ESF#13-12).

In Emergency Support Function 14, "Long-Term Community Recovery Annex," the Department of Commerce's National Oceanic and Atmospheric Administration (NOAA) is tasked to provide "natural hazard vulnerability analysis, provides assistance on coastal zone management and building community resilience, supplies geospatial technology (e.g., Geographic Information System, or GIS) assistance and coastal inundation information, performs ecosystem and damage assessments" (Federal Emergency Management Agency (FEMA), 2008, ESF#14-6).

Explicit mention of GIS as a way to link assets and services across various response and other mission areas, as discussed in the NRF, creates a unique opportunity for GIS to be further integrated into other aspects of disaster management. The explicit mention of GIS in a wide variety of ESFs helps to demonstrate this point with GIS being mentioned for traditional disaster management activities such as planning, but also more nontraditional and perhaps even unexpected activities such as preserving natural and cultural resources, helping to maintain food safety, incorporation into the activities of well-known groups such as NASA and disaster mitigation activities, which will be discussed in Chapter 7. Ideally, over time, as the value of GIS continues to be proven, GIS and related spatial activities will continue to get explicit mention in other ESF activities.

The next section discusses specific ideas around the use of GIS for disaster response activities.

GEOGRAPHICAL ASPECTS OF SITUATION AWARENESS

As you may recall from Chapter 1 and the fundamental "who, what, where, when, why and how" questions GIS can answer, during disaster response, the "where" and "what" aspects are the most important. In this regard, disaster response is the one disaster management cycle phase in which the idea of *situation awareness* you have heard repeatedly throughout this book is perhaps the most relevant in terms of GIS serving as a situation awareness support mechanism. As you first saw in Chapter 1 and Figure 1.3 (the historical military map from World War II of the Normandy Beach landings), maps have a long tradition as the physical embodiment of situation awareness. The parallels between military and disaster management activities are close in this regard. For example, disaster management decision makers need to be constantly aware of the situation they are dealing with in a number of different dimensions such as locations of response personnel, areas to evacuate, disaster victims, and location of relief supplies much like a general needs to be kept aware of troop locations and supplies to keep troops operational. Furthermore, the time-sensitive nature of disaster response also dictates that maps and other spatial support devices are capable of keeping up with the time-sensitive nature of the disaster response situation and be able to generate disaster response GIS products that are readable and consumable for relevant audiences that need them. As will be discussed in the "Spatial Data Deluge" section of this chapter, this is in fact a very challenging aspect of disaster information management. The following sections discuss items relevant to the use of GIS and maps for supporting disaster response situation awareness.

Maps and Emergency Operation Centers

A common image people may have about disaster response activities is that of a large room containing multiple information screens with various people working to respond to a disaster. This is, in fact, is called an Emergency Operations Center or EOC (Figure 6.2).

There are several interesting things to make note of in Figure 6.2, which shows an EOC activated during Hurricane Ike in San Antonio, Texas, in 2008. First, in the foreground, make note of the people who are sitting around computers with a sign that says LOGISTICS. This is an example of a section of the Incident Command System, or ICS, which you learned about in Chapter 4 being activated during an incident. Second, make note of the other people sitting around at various desks and computers in the background that are most likely from other ICS sections and various local authorities that have been activated because of the incident. Finally, and most importantly for this book, make note of the two large monitors that are in the top center of the image displaying real-time maps

Figure 6.2   An Emergency Operations Center (EOC). (FEMA photo by Jocelyn Augustino.)

of the hurricane's progress. This is an excellent example of the use of maps to provide situation awareness, and in this case, a common operating picture (a term discussed in Chapter 5) to support overall group coordination and cooperation (Breen and Parrish, 2013). Also make note how the maps are being juxtaposed with national news media reports used to keep responders aware of the situation and that specific contents on individual monitors can be changed as the situation evolves. An EOC will typically have a dedicated GIS supervisor or analyst who is tasked with providing real-time updates on the large screens found in an EOC.

GIS and Disaster Warnings

Straddling the line between disaster planning and disaster response, GIS and subsequent map products can play an important role in providing the public with spatial representations of disaster warnings that can provide the initial disaster response situation awareness. As seen in Figure 6.1, maps are commonly used for displaying oncoming hazards like an approaching hurricane.

Furthermore, in the age of smartphones, tablet computers, and other technologies, people are increasingly being connected together in real time with vast amounts of data from vast numbers of sources such as real-time weather information, real-time earthquake information, news feeds, and social media posts. In this regard, mobile apps (applications) such as Disaster Alert created by the Pacific Disaster Center or map-based disaster warning services such as the Interior Geospatial Emergency Management System (IGEMS; http://igems.doi.gov/) and others are proving to be valuable sources of automated, map-based disaster warning information.

In addition to the use of GIS-generated maps as communication mediums for disaster warning, GIS is also an important analytical tool for when to actually issue a disaster warning or evacuation order. Issuing of disaster evacuation orders can be a very delicate matter as issuing the order at the wrong time, when the threat from the disaster is actually not very high, can lead to the problem that citizens will not heed future warnings given because they will have memories of the past warnings not actually being dangerous, or "crying wolf" scenarios. GIS can be used to determine the exact moment at which the disaster warning or evacuation order should be issued so that citizens have enough time to evacuate, but not so soon as to diminish future trust in warnings. Cova et al. (2005) presented work in this regard in terms of developing a spatiotemporal model to determine the spread of wildfires and the specific time point at which an evacuation order should be issued.

Now that you have ideas about the geographical aspects of situation awareness, it is important to discuss a common problem in very large disasters, that of spatial data deluge.

SPATIAL DATA DELUGE

Spatial data deluge is the idea that the sheer, overall volume of spatially referenced data coming into the disaster response coordination and decision mechanisms is so overwhelming that the data simply cannot be processed and acted upon to be of actionable value. This type of problem was identified earlier in this book when you read the interview by Alan Leidner in Chapter 4 where he discussed the problem of overwhelming amounts of data that were generated as Hurricane Sandy situation unfolded in 2012. However, spatial data deluge has routinely been a problem in large disasters ranging anywhere from the Indian Ocean tsunami of 2004 all the way to the most recent disasters at the time this book was written such as the Philippines Yolanda event of late 2013. Processing, analyzing, curating, making sense of, and communicating relevant information derived from large amounts of data is in fact an ongoing research challenge for GIS for disaster management. For example, at the time this book was written, social media such as Twitter have received significant attention as a source of disaster situation information that is massive in terms of the overall volume of tweets that are generated with much attention dedicated to developing computational and other methodological approaches as for processing large tweet volumes to inform emergency situations (Paul and Dredze, 2011; Verma et al., 2011).

Although handling large volumes of spatial data is an ongoing research challenge, the following is some practical advice that can be followed to help deal with the problem of spatial data deluge using existing commercial and open-source GIS tools were discussed in Chapter 3. In general, the overall strategy for using these techniques is to aggregate data by identifying potential clusters of data that are of interest in terms of situation awareness or provide evidence for broader spatial decision support.

Thematic Maps

Perhaps the oldest examples of handling geographic "big data," thematic maps are inherently designed to show aggregated data using statistics to define data class breaks. As you saw in Chapter 2, choropleth maps, proportional symbol maps, and isarithmic maps are used to aggregate data and communicate data using spatial representations in map-based forms. However, care should be used by selecting the appropriate display technique that matches the data being displayed.

Spatial Statistics

Spatial statistics are a collection of statistical techniques designed to investigate how the underlying *spatial* nature of phenomena may or may not influence the statistics generated (for which traditional statistics cannot account). Spatial statistics are in fact relevant to many phases of the disaster cycle such as disaster planning, response, and recovery, for example, tracking the outbreaks of a disease, 911 call patterns, or analyzing large volumes of tweets ben generated during a disaster (Waller and Gotway, 2004; Sizov, 2010; Cutter and Finch, 2008; MacEachren et al., 2011). Although spatial statistics are a vast topic unto themselves, the following is one spatial statistic technique that is commonly used for handling spatial data at deluge and is known as *hot spot* or *heat* mapping. Note that the "Hot Spot Mapping" section assumes the reader is familiar with basic statistical concepts such as confidence intervals and Z-scores; for details on these topics, see Abramovich and Ritov (2013).

Hot Spot Mapping

One hot spot mapping technique known as the Gi^* statistic is designed to find statistically high (*hot*) or low (*cold*) value clusters. When calculating, every feature in a dataset is compared with neighboring features within a specified distance to determine the extent with which each feature is surrounded by neighbor features of similar low or high values (Mitchell, 2009).

The Gi* statistic works in the following manner: First, a neighborhood is defined using either a set distance (which is based on prior knowledge of features and their behaviors) or adjacent features (Mitchell, 2009). As disaster management examples, hot spot detection based on a set distance might be used in a case of determining how far people are willing to travel to a temporary evacuation shelter set up during a disaster response damage assessment or to find concentrations of vulnerable populations by looking at attribute values of the adjacent census tract locations (Figure 6.3).

The size of the distance used will determine the size of the cluster created. For example, the greater the distance, the larger the clusters. Distance can be spatially conceptualized using a variety of approaches, such as straight line distance, Euclidean distance, or travel time (for further details, see Mitchell, 2009).

Second, for calculating the Gi* statistic, "the GIS sums the values of the neighbors and divides by the sum of the values of all the features in the study area" (Mitchell, 2009, 176).

Furthermore, Gi* uses a binary weight for calculation (Mitchell, 2009). For example, in a case where we are interested in finding clusters of vulnerable populations in census

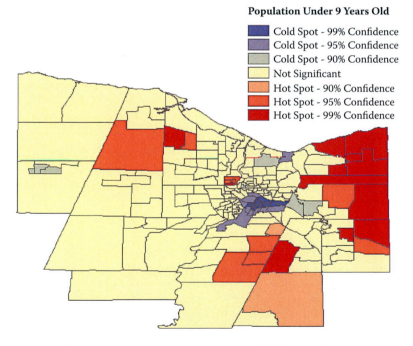

Population Under 9 Years Old

Cold Spot - 99% Confidence
Cold Spot - 95% Confidence
Cold Spot - 90% Confidence
Not Significant
Hot Spot - 90% Confidence
Hot Spot - 95% Confidence
Hot Spot - 99% Confidence

Figure 6.3 Hot spot detection based on adjacent edge calculation of census tracts within a US county containing individuals under 9 years of age. Note the distinctive "cold" spot in the center of this graphic (which is an urban area) that shows a clear pattern that young people are not concentrated in this location but are generally more located on the periphery. (Map by Brian Tomaszewski.)

tracts, the statistic will look at attribute values such as the count of vulnerable people living in the census tract and will multiply that value by either one (if it is a neighbor) or zero (if not a neighbor) to ensure that neighbors are included in the calculation.

One version of the Gi* formula from Mitchell (2009, 176) is as follows:

$$G_i^*(d) = \frac{\sum\limits_j W_{ij}(d) X_j}{\sum\limits_j X_j}$$

where:

$G_i^*(d)$ = the G_i^* for a feature i at a distance d

$\sum W_{ij}(d) X_j$ = The value of each neighbor X_j is multiplied by a binary weight for the target (or current feature being examined in relation to other features) and neighbor pair (W_{ij}) and the results are summed $(\sum)$.

This sum is then divided by the sum of all neighbor values $(\sum X_j)$ or all dataset features.

The Gi* statistic results are interpreted in that feature groups with high Gi* values represent hot spots or feature clusters with high attribute value, feature groups with low Gi* values are also clusters of cold spots, Gi* values around 0 do not have low or high concentrations and this situation occurs when values surrounding are target feature are either near the mean or a mix of low and high values (Mitchell, 2009).

To test for statistical significance, the following formula, based on (Mitchell, 2009, 178) is used:

Gi* Z-score:

$$Z(G_i^*) = \frac{G_i^* - E(G_i^*)}{\sqrt{Var(G_i^*)}}$$

where $Z(Gi^*)$ is a Z-score for the Gi*, which is calculated by subtracting an expected Gi* for a feature given (and based on a random spatial distribution) from the actual calculated Gi* value (discussed previously) or in formula notation : $G_i^* - E(G_i^*)$, the difference of which is then divided by the variance square root for all study area features or in formula notation, : $\sqrt{Var}(G_i^*)$ (Mitchell, 2009).

The expected random distribution G_i^* formula based on (Mitchell, 2009, 178) is

$$E(G_i^*) = \frac{\sum W_{ij}(d)}{n-1}$$

where:

$E(G_i^*)$ is the expected Gi* value

$\sum W_{ij}(d)$ is the sum of weights at given distance, which is divided by $n-1$ or the overall number of features minus one.

197

Z-scores are also calculated at a specific distance where, like the Gi*, score, high Z-score values are indicative of high attribute neighbor values, low Z-score values indicate low attribute values, and near-zero Z-score indicate no apparent similar value concentrations (Mitchell, 2009) (Figure 6.4).

Z-score statistical significance is then determined by comparing the Z-score to value ranges within a specific confidence interval (as seen previously) in Figure 6.3 (Mitchell, 2009).

Graduated color maps can then be used to visually represent Gi* statistic results in terms of the Gi* values themselves or the Z-score to show statistically significant areas and cluster patterns in general (Mitchell, 2009).

It also important to be aware of factors that influence the Gi* results. For example, since the Gi* is calculated based on feature adjacency or specific distances, features at the study area edge will have fewer neighbors and will thus be skewed because there are fewer

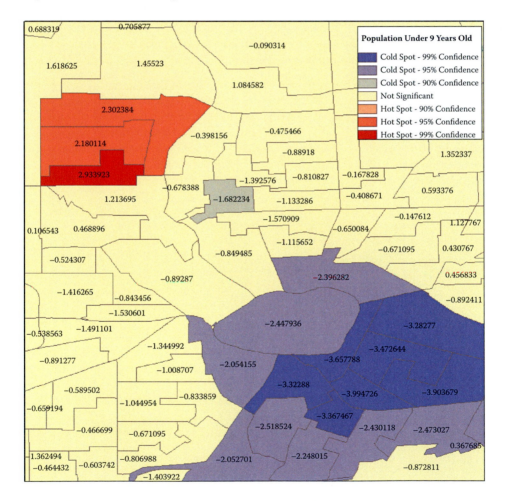

Figure 6.4 Close-up of Figure 6.3 showing Z-scores per individual census tract, and their assignment to confidence intervals. (Map by Brian Tomaszewski.)

overall features to calculate, and this smaller set of features may have greater importance in the calculations (Mitchell, 2009). Other issues to account for include small samples (<30), which may affect the results due to outliers or global patterns that may make localized clusters less obvious (Getis and Ord, 1996; Mitchell, 2009).

Density Mapping

Density mapping is the idea of defining areas based on the density or count of features. In a disaster response context, this can be a useful technique for understanding where patterns are emerging from large data input volumes, for example, frequency of 911 calls or instances of social media related to an ongoing disaster (Figure 6.5).

The top left of Figure 6.5 shows the overall number of tweets generated during Hurricane Sandy (2012) in Manhattan. As can be seen in the top left figure, the sheer

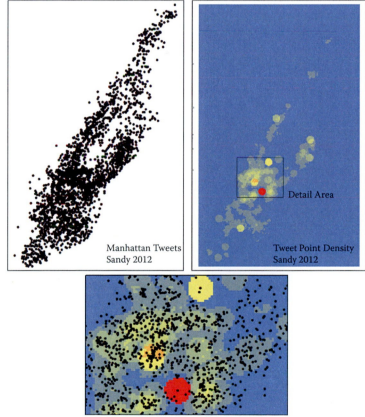

Figure 6.5 Point density mapping of tweet locations in Manhattan during Hurricane Sandy in 2012. Colors represent red (higher) to blue (lower) tweet volume. Note that multiple tweets might be located on a single point as per the nature of Twitter and thus create higher density clusters (as is the case with the red cluster).

overall number of tweets make it very difficult to discern any sort of useful spatial pattern. The figure on the right side of Figure 6.5 shows the point density calculation generated based on the count of tweets using the ArcMap point density tool. The output of this tool is a raster surface where the grid cells in the raster represent point densities. In the bottom middle of Figure 6.5, the tweet locations and density raster surface are combined and zoomed in on a detailed area so you can see the specific relationship between tweet locations and the density raster grid. Note that outputs created by tools like point density and others that calculate densities are subject to various tool input parameter settings. For example, the density patterns shown in Figure 6.5 on the top right side are calculated based on the defining a search "window," which in the case of the right side of Figure 6.5, is a circle. This means that when the density tool calculates point densities, a circle within a user-defined radius is used to examine data in terms of the number of points falling inside a circle, which in turn are used to calculate the density. These calculations are then used as inputs for creating an output raster grid representation of density, which can be stylized using intuitive colors ranging from blue for low density to red for high density. Changing the size of the search window will have the effect of changing how the density is represented. For example, the larger the search window based on a circle with a bigger radius, the coarser the representation of density as opposed to a smaller search window based on a smaller circle radius, which will create a finer density representation. Determining the exact size of a search window should be based on the question being asked, the extent of the study area, and the nature of the data itself in terms of its spatial distribution. For more discussion on this topic, see Silverman (1986).

Real-Time GIS

The time-sensitive, critical nature of disaster response makes *real-time GIS* for handling spatial data deluge very important. Real-time GIS the idea that data and inputs with a spatial reference can be incorporated into a GIS for decision making as soon as the data itself has been created, for example, tracking the location of response vehicles using a global positioning system (GPS) receiver inside a vehicle, monitoring inputs sent from field data assessment teams using mobile device apps, or even real-time or near-real-time data streams of imagery being collected and spatially referenced from an airplanes or drones flying over a disaster area (van Aardt et al., 2011).

An excellent example of the newly emerging real-time GIS technology very relevant for disaster response is Esri's GeoEvent processor. The GeoEvent processor technology works with the ArcGIS server to connect to a variety of sensors such as social media feeds and global positioning system receivers to then collect data in real time, process and filter the data based on user needs such as filtering out particular disaster response units such as ambulances arriving at a hospital versus police cars, and then uses the input data to push alerts and other notifications to relevant parties. For example, when an ambulance is detected to be within 1 mile of a hospital, Short Message Service (SMS) messages can be sent to hospital officials to alert them about the incoming ambulance. Figure 6.6 is a general example of how the GeoEvent processor technology works.

Figure 6.6 Example of Esri's GeoEvent Processor tool. In this example, an alert is triggered when an ambulance has crossed what is known as a *Geo fence* or a predefined area of interest that is used to communicate to an asset, which in this example is a hospital. As can be seen in the bottom right of the figure, personnel at the hospital receive an SMS message when the ambulance is arriving. (Copyright © 2014 Esri, ArcGIS, ArcMap. All rights reserved. Used with permission.)

DISASTER RESPONSE GIS PRODUCTS

In addition to the importance of dealing with spatial data deluge as described previously, it is important to consider disaster response GIS products. Figure 6.7 is a disaster response GIS product development framework.

Ideas in the framework tie in with many other ideas you have learned about so far in this book. For example, at the top of Figure 6.7 is reference data collected during the planning phase (as discussed in Chapter 5). Referenced data then provide essential context and ultimately contextualize a variety of situation inputs derived from a variety of input sensors. Situation inputs themselves may then be processed and analyzed using a variety of techniques such as clustering pattern analysis discussed in this chapter and other GIS analytical techniques. Data processing and analytics will also be influenced by factors such as the specific software used, the computing hardware power available, the skill level of the GIS person, and the situation complexity. After being processed and analyzed, specific products then can be developed. Development of specific products is also influenced by factors that must be accounted for.

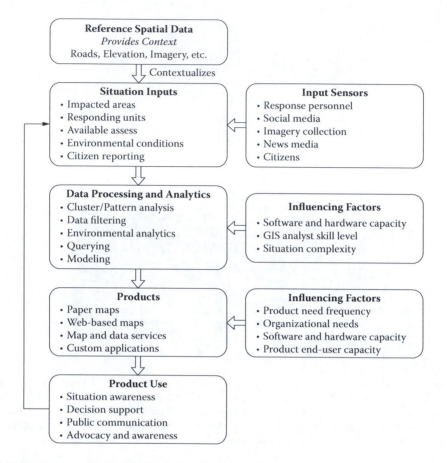

Figure 6.7   A disaster response GIS product framework.

For example, it is very important to keep in mind the capacities of people who are the actual responders and consumers of GIS products created during a disaster response. Paper-based maps are still a very important medium by which disaster response information is communicated in map-based formats. Paper maps offer the advantage of being extremely portable, have no power requirements, can be easily annotated with hand drawings, and require absolutely no computing skill to use, which can be very important when dealing with a wide range of disaster management professionals with varied backgrounds. Although web-based mapping presentation continues to proliferate, it is important to keep in mind that web-based tools and representations will be of little use if there is no Internet connection or hardware that can read and view web-based maps. Smaller screen sizes of maps presented on smartphones and other devices may be difficult to read, especially if there's a lot of data being presented, and again, will be generally subject to the computing skill level of the people trying to use the web-based representation. However, web-based maps do have significant advantages because if the Internet is available, they can be disseminated freely and widely. Furthermore, web-based map presentation can have the advantage that end users do not have to have purchase expensive software licenses associated with commercial software for viewing maps, as often is the case in desktop computing environments.

Another important factor that influences the development of GIS disaster response products is the frequency with which products are required, and the specific organizational needs so that the products themselves are relevant. For example, disaster responders focusing on cultural heritage issues (as discussed previously) will need maps focusing on historic and culture themes and not necessarily on other thematic areas such a shelter locations or locations of command headquarters.

When specific products are developed, they can have a variety of uses such as situation awareness, decision support, public communication, and general advocacy in awareness of disaster situations in cases where external donations and aid may be required (O'Connor, 2008; Tomaszewski and Czárán, 2009). Actual GIS disaster response products can also serve as situation inputs to decisions and other actions that are taken based on the products and guide new product development.

Online Disaster Response Geographic Data Streams

A different type of disaster response GIS project is online disaster response and geographic data streams. This is largely driven by the fact that the high public visibility of disaster response activities has increasingly drawn the interest of large information technology and data companies such as Google, Microsoft, and Esri in recent years as part of their philanthropic and public outreach activities. These companies provide disaster response–relevant data that they collect, which can be used to complement official government data sources. For example, the Google Crisis Response team (https://www.google.org/crisisresponse/) often creates and freely disseminates crisis-relevant data in Google data formats such as KML and Google Maps–based applications that can be used as inputs for development of other disaster response products. Google will also develop custom applications (https://www.google.org/crisisresponse/resources.html) relevant to disaster response such as Google Alerts and Google Person Finder in addition to promoting their general suite of online collaborative tools such as Google Documents, Google Spreadsheets, and Google Maps, which are widely used during global disaster response due to their ease of use and accessibility by a wide range of people.

Similar to Google, Esri also provides a disaster response service closely tailored to their software and data structures in the form of public awareness maps, dataset downloads, and free technical assistance and access to their software (if approved by Esri) for use during disasters (Figure 6.8; see the Resources section for more details).

GIS and Damage Assessment

Another common activity during a disaster response, that leads to the creation of GIS disaster response products, is damage assessment. Damage assessment is the idea of collecting data on the level of destruction, causalities, and other factors during a disaster to gauge the level of response and recovery needed, for example, using a tax parcel layer to assess the level of damage done to buildings after a flood (Figure 6.9).

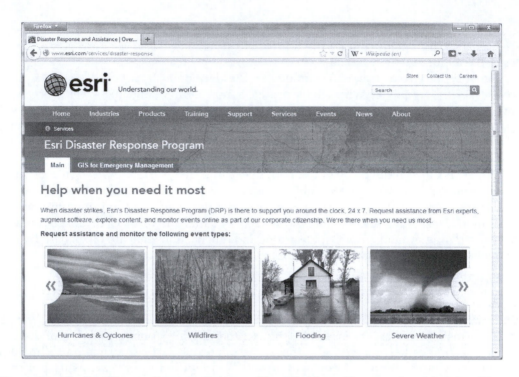

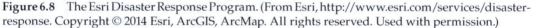

Figure 6.8 The Esri Disaster Response Program. (From Esri, http://www.esri.com/services/disaster-response. Copyright © 2014 Esri, ArcGIS, ArcMap. All rights reserved. Used with permission.)

Damage assessments are often one of the first activities done during a disaster response to judge the level of severity of the disaster in terms of potential assistance requests. Damage assessment is also an excellent example of the use of GIS in field-based or mobile capacity such as going out to a disaster zone with a small computer, smartphone, or any other mobile device that can capture field data (often with the use of the global positioning system receiver and the built-in camera and video recording features of the device) to conduct damage assessments. The following section briefly describe two GIS field data collection tools. Although these tools are not specifically designed for disaster response activities, their capabilities can easily be extended for use in damage assessment or any other GIS disaster management need.

Field Data Collection and Mobile GIS
Commercial Technology: Esri's ArcPad
ArcPad software by Esri (http://www.esri.com/software/arcgis/arcpad) is a well-established, mature software environment for mobile field data collection (Figure 6.10).

As seen in Figure 6.10, the ArcPad environment is specifically designed for use in small-screen environments, and can utilize existing GIS data for field collection of features. Make note in Figure 6.10 of the minimal number of icons that are used and the small

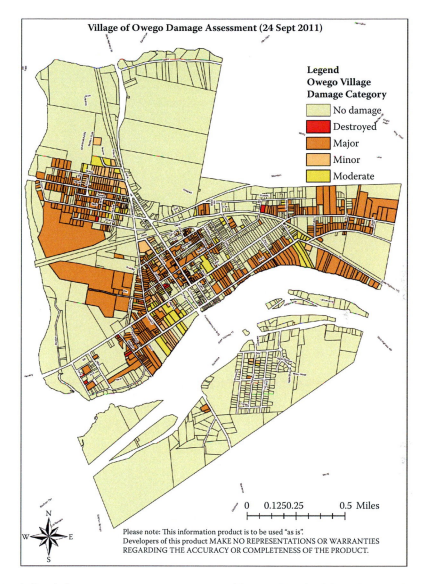

Figure 6.9   A flood damage assessment map created by marking building damage levels on a tax map and rendered using variable color to indicate damage levels. This map was created during a 2011 flooding incident in Owego, New York. (Map by Brian Tomaszewski.)

data entry screen. All of these features can be customized in ArcPad depending on user needs. Furthermore, it is designed to be fully integrated with a broader GIS infrastructure such as synchronized data editing in integration with ArcGIS server feature services. For a case study on using ArcPad for disaster response assessments, see Environmental Systems Research Institute (2005).

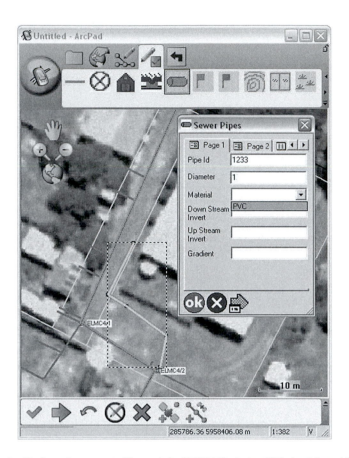

Figure 6.10 The ArcPad environment. (Copyright © 2014 Esri, ArcGIS, ArcMap. All rights reserved. Used with permission.)

Open-Source Technology: ODK Collect

OpenDataKit (ODK) Collect (http://opendatakit.org/use/collect/) is an open-source software environment specifically designed to run in Android operating system–based environments. It is part of the broader ODK collection of software (http://opendatakit.org/) that includes ODK Aggregate, which is a server-based environment designed for gathering sharing, storing, and analyzing data captured with ODK Collect. ODK Collect and other associated tools in the ODK software suite are based heavily on Google technology such as exporting geographic data collected by ODK Collect into the KML format. ODK Collect uses the concept of forms, which are simple interfaces that can be designed using XML to collect data in the field and customizable for whatever data needs are required (Figures 6.11a and 6.11b).

Figures 6.11a and 6.11b are screen shots of the ODK Collect main menu and Geo Tagger v2 form inside the ODK Collect environment running on a tablet device. The Geo Tagger v2 form (Figure 6.11b) walks users through the collection of point-based data in a very easy-to-use format such as using very large buttons as seen in the image.

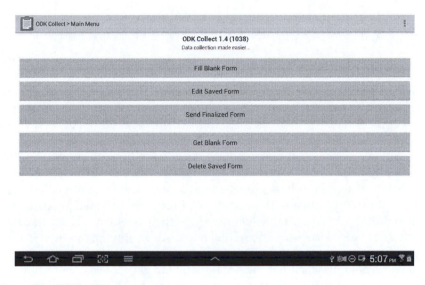

Figure 6.11a   The ODK Collect main menu.

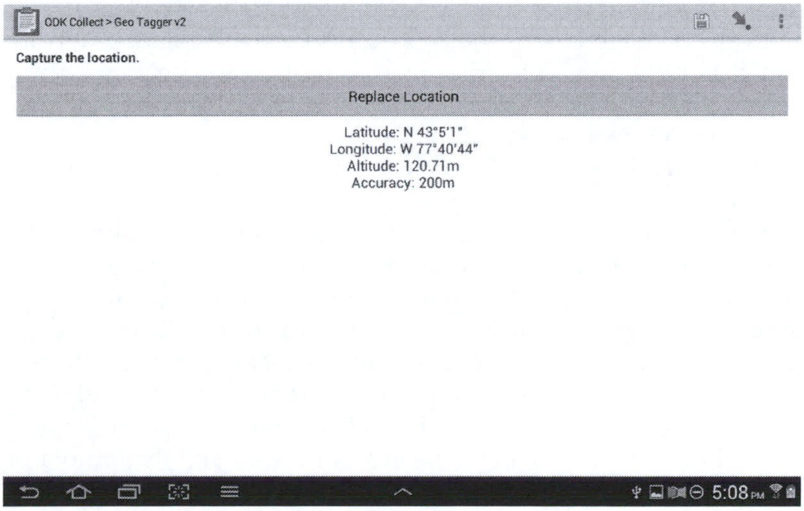

Figure 6.11b   The ODK Collect Geo Tagger v2 form for collecting spatial data.

The open-source nature of ODK Collect and the fact that it is designed to run on the Android operating system (which is itself an open source) makes ODK Collect a compelling technology choice for disaster assessment and any other field-based operations when financial constraints are potential factors in using commercial solutions. For a case study on the use of ODK Collect for disaster assessment activities, see Anokwa (2011).

PUBLIC AND DISASTER RESPONSE MAPPING—
CRISIS MAPPING AND CITIZEN REPORTING

Increasingly, nonprofessional disaster management people are becoming involved in disaster response through the use of various crisis mapping techniques that have been enabled through greater access to mapping technology in general—ideas first mentioned in Chapter 1. Disaster response is the disaster management phase than most often gets the attention of people interested in crisis mapping. This is especially true in the case of very large disasters such as the Haiti earthquake of 2010 where the massive scope, media exposure, and overall international attention garnered a great outpouring of good will from people who wanted to help with the situation. Additionally, victims of disasters can in fact be responders to a disaster themselves. For example, people who were impacted by Hurricane Sandy of 2012 were very active with hosting need requests on crisis maps in using other social media outlets such as Twitter to communicate their personal situations and needs.

Like any other GIS or map-based approach to disaster management, it is important to recognize the benefits and drawbacks of involving the public in disaster response mapping. The benefits lie in the fact that the public can help gather large amounts of data from a wide variety of people to fill information gaps, as you saw the example in Chapter 1 with the use of crisis mapping techniques to fill information gaps in the Syria civil war due to media blockage by various government agencies. Drawbacks of crisis mapping, however, stem from reliability and verifiability of information that is collected. Although steps can be taken to alleviate these issues, such as verifying and checking information before it is posted to a publicly accessible crisis map, if crisis mapping is collecting data from unknown people, any data that is collected should be critically examined before using the data as decision-making inputs.

If you are interested in exploring crisis mapping and map-based crowdsourcing tools, Crowd Map (https://crowdmap.com/welcome) is a great place to start (Figure 6.12).

Crowdmap is based on Ushahidi technology first discussed in Chapter 1. Crowdmap, however, provides an advantage in that the underlying technology is hosted in the cloud by Crowdmap, thus requiring very limited technical knowledge and resources to start using a Crowdmap instance. By filling out a simple form, one can start using Crowdmap almost instantly. If you are comfortable using advanced technology such as open-source web servers, databases, and programming environments such as PHP, then the full Ushahidi platform (http://www.ushahidi.com/) is another option for getting started with crisis mapping that can provide more customization and flexibility options.

CHAPTER SUMMARY

In this chapter you learned about Geographic Information Systems for disaster response. First, you learned about disaster response policy in the United States and how it relates to GIS. Specifically, you learned about the NRF and how GIS is applicable to a wide range of emergency support functions such as planning, environmental and agricultural resources, and even cultural heritage protection. Next, you learned about the geographical aspects of situation awareness where GIS plays its most important role during

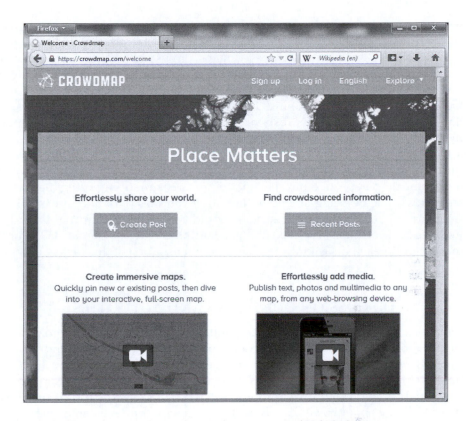

Figure 6.12   The Crowdmap website. (Screenshot courtesy of Ushahidi.)

disaster response. In this regard, you saw how maps and related GIS functionality are central to the functioning of an emergency operations center and the importance of maps and GIS as disaster warning devices.

Next, the chapter discussed various techniques for handling spatial data deluge, or the massive volume of data that is generated during a disaster response. Specifically, you were given an introduction to hot spot mapping techniques, which can help determine statistically significant clusters of data, and you were shown density mapping techniques that can help determine where spatial data density patterns occur. Both of these techniques are vast topics unto themselves, and you are strongly encouraged to review some of the items in the resources section of this chapter to learn more about these topics in general, and then think about how those ideas can be used for your own disaster response activities.

The topic of a real-time GIS was then presented as another idea to consider for handling spatial data deluge. In particular, you learned about Esri's GeoEvent Processor tool, which is specifically designed to handle real-time, spatial reference data streams such as social media feeds and GPS receiver data for rapid analysis and decision making.

The chapter then discussed disaster response GIS product development, which is very important during a disaster response for decision support and maintaining situation awareness. The framework presented in that section should be used as a loose guideline for things to think about as you develop your own GIS disaster products, whether they are hard-copy base maps, interactive web-based maps, or custom software applications. You were also given some perspectives on the broader corporate IT world and disaster response and how companies like Google and Esri offer philanthropic services related to crisis response that can provide opportunities for free GIS data and software during disasters.

Related to the topic of disaster response GIS products, you were given some perspectives on the use of mobile GIS capabilities for field data collection and were shown two specific technologies for this purpose. Given the high visibility that disaster response has in the public eye, the chapter then discussed the role of the public in disaster response mapping from the perspective of crisis mapping. Crisis mapping offers opportunities for disaster response by being able to gather large volumes of data that can fill in data and information gaps. Conversely, caution should be exercised that the data collected and is reliable and verifiable. The next chapter continues with the theme of GIS for disaster management in each disaster cycle phase and focuses on GIS for disaster recovery.

DISCUSSION QUESTIONS AND ACTIVITIES

1. Refer back to the 14 core response capabilities outlined in Table 6.1. Look through each of those core capabilities, and come up with specific GIS tasks and scenarios that you could imagine being used to support the specific capability. For example, developing a transportation GIS database that could be used to support core functionality 4 (critical transportation systems). As always, you are encouraged to think spatially about how GIS can be used as a support device to engage spatial aspects of these core response capabilities. Use the Internet or other research tools at your disposal to find specific case studies of GIS applications for disaster response that might fit the core response capability that you are investigating to help ground your ideas in a real-world context, or perhaps form the basis for new original research you might conduct.
2. Use the Internet to find information about your local emergency management services, and see if you can determine if your town, county, city, or state has their own EOC. Contact your local officials to see if they would give you a tour of the EOC and find out how they use GIS to support EOC activities.
3. Besides the example of wildfires given in the GIS and disaster warnings section, what other kinds of specific natural hazards can you think of that would have a sensitive evacuation order trigger point? Based on the natural hazard you think of, what kind of GIS model might you develop to determine when exactly to issue an effective disaster warning or evacuation order that people will believe and trust?
4. Think of any topic that is spatial in nature and in which you would use crowd mapping to increase understanding. Visit the Crowdmap website and create an instance of Crowdmap to investigate your topic. Working with your classmates, add data to your Crowdmap instance. After using Crowdmap for a while, what do you like best about it? What didn't you like about it? Could you see using Crowdmap during an actual disaster? Why or why not?

RESOURCES NOTES

See http://www.fema.gov/media-library-data/20130726-1918-25045-9979/gissupervi-sor.pdf as an example of a GIS for disaster management job position that requires (1) ICS–EOC interface and (2) EOC management and operations training.

For more information on the Disaster Alert app, see Google, https://play.google.com/store/apps/details?id=disasterAlert.PDC.

For more information on the general topic of spatial statistics, see the Esri Press website for the book *The ESRI Guide to GIS Analysis, Volume 2,* http://esripress.esri.com/display/index.cfm?fuseaction=display&websiteID=86, a gentle introduction to the topic of using Esri tools, and http://esripress.esri.com/display/index.cfm?fuseaction=display&websiteID=194, for a description of the book *Spatial Statistical Data Analysis for GIS Users,* an advanced introduction to the topic of using Esri tools.

For information on spatial statistics and modeling, see the Springer website, http://www.springer.com/statistics/statistical+theory+and+methods/book/978-0-387-92256-0, for a description of the book *Spatial Statistics and Modeling,* and the R Project website for information on using the R open-source statistical package (http://www.r-project.org/).

For some practical tutorials on how to do hot spot mapping using Esri tools, see http://resources.esri.com/help/9.3/arcgisengine/java/gp_toolref/spatial_statistics_toolbox/spatial_statistics_toolbox_sample_applications.htm.

Hot Spot Analysis reference in Esri, see http://help.arcgis.com/en/arcgisdesktop/10.0/help/index.html#/Hot_Spot_Analysis_Getis_Ord_Gi/005p00000010000000/.

For Google versions of density and heat mapping tools, see https://support.google.com/fusiontables/answer/1152262?hl=en.

Esri Point Density Tool reference, see http://resources.arcgis.com/en/help/main/10.1/index.html#/Point_Density/009z0000000v000000/.

For more information on Esri Emergency software assistance, see http://www.esri.com/apps/company/assist/index.cfm?eventID=121.

See the Ushahidi blog at http://blog.ushahidi.com/2012/10/29/hurricane-sandy-in-maps/ as a crisis map example from Hurricane Sandy in 2012.

See http://www.esri.com/software/arcgis/smartphones/collector-app for more Esri apps such as the Collector for ArcGIS app designed to collect field data on smartphones and tablets.

REFERENCES

Abramovich, Felix, and Ya'acov Ritov. 2013. *Statistical Theory: A Concise Introduction.* Boca Raton, FL: CRC Press.

Anokwa, Yaw. 2011. "Wisconsin using ODK for natural disaster assessments," OpenDataKit, October 14, http://opendatakit.org/2011/10/wisconsin-using-odk-for-natural-disaster-assessments/ (accessed March 8, 2014).

Breen, Joseph J., and David R. Parrish. 2013. "GIS in emergency management cultures: An empirical approach to understanding inter- and intra-agency communication during emergencies." *Journal of Homeland Security and Emergency Management* 10 (2).

Cova, Thomas J., Philip E. Dennison, Tae H. Kim, and Max A. Moritz. 2005. "Setting wildfire evacuation trigger points using fire spread modeling and GIS." *Transactions in GIS* 9 (4):603–617.

Cutter, Susan L., and Christina Finch. 2008. "Temporal and spatial changes in social vulnerability to natural hazards." *Proceedings of the National Academy of Sciences* 105 (7):2301-2306.

Environmental Systems Research Institute. 2005. *North Carolina Division of Public Health: Mobile GIS Speeds Disaster Relief*, Esri, http://www.esri.com/software/arcgis/arcpad/~/media/Files/Pdfs/library/casestudies/nchealth.pdf (accessed March 8, 2014).

Federal Emergency Management Agency (FEMA). 2008. Emergency Support Function #5 – Emergency Management Annex. http://www.fema.gov/media-library-data/20130726-1913-25045-2444/final_esf_5_information_and_planning_20130501.pdf

Federal Emergency Management Agency (FEMA). 2008. ESF Annexes Introduction. http://www.fema.gov/media-library-data/20130726-1825-25045-0604/emergency_support_function_annexes_introduction_2008_.pdf

Federal Emergency Management Agency (FEMA). 2013. Information Sheet: National Response Framework. http://www.fema.gov/media-library-data/20130726-1914-25045-6465/final_informationsheet_response_framework_20130501.pdf

Getis, Arthur, and J. Keith Ord. 1996. "Local spatial statistics: An overview. " In *Spatial Analysis: Modelling in a GIS Environment*, edited by P. Longley and M. Batty. New York: Wiley.

MacEachren, Alan M., Anuj Jaiswal, Anthony C. Robinson, Scott Pezanowski, Alexander Savelyev, Prasenjit Mitra, Xiao Zhang, and Justine Blanford. 2011. "SensePlace2: GeoTwitter analytics support for situational awareness." Paper read at IEEE Conference on Visual Analytics Science and Technology (IEEE VAST), at Providence, RI.

Mitchell, Andrew. 2009. *The Esri Guide to GIS Analysis, Volume 2: Spatial Measurements and Statistics*. Redlands, CA: Esri Press.

O'Connor, Sean. 2008. *Maps for Advocacy: An Introduction to Geographical Mapping Techniques*, Berlin: Tactical Technology Collective.

Paul, Michael J., and Mark Dredze. 2011. "You are what you tweet: Analyzing Twitter for public health." Paper read at International AAAI Conference on Weblogs and Social Media (ICWSM).

Phillips, Brenda D., David M. Neal, and Gary R. Webb. 2012. *Introduction to Emergency Management*. Boca Raton, FL: CRC Press.

Silverman, B.W. 1986. *Density Estimation for Statistics and Data Analysis*. New York: Chapman and Hall.

Sizov, Sergej. 2010. "Geofolk: Latent spatial semantics in web 2.0 social media." Paper read at Proceedings of the Third ACM International Conference on Web Search and Data Mining.

Tomaszewski, Brian, and Lóránt Czárán. 2009. "Geographically visualizing Consolidated Appeal Process (CAP) information." Paper read at Proceedings of the 6th International Information Systems for Crisis Response and Management (ISCRAM) Conference, at Gothenburg, Sweden.

United States Department of Homeland Security. 2013. *National Response Framework*, 2nd edition, FEMA, http://www.fema.gov/national-response-framework (accessed June 7, 2014).

van Aardt, Jan, Donald McKeown, Jason Faulring, Nina Raqueño, May Casterline, Chris Renschler, Ronald Eguchi, David Messinger, Robert Krzaczek, and Steve Cavillia. 2011. "Geospatial disaster response during the Haiti earthquake: A case study spanning airborne deployment, data collection, transfer, processing, and dissemination." *Photogrammetric Engineering and Remote Sensing* 77 (9):943–952.

Verma, Sudha, Sarah Vieweg, William J. Corvey, Leysia Palen, James H. Martin, Martha Palmer, Aaron Schram, and Kenneth Mark Anderson. 2011. "Natural language processing to the rescue? Extracting "situational awareness" tweets during mass emergency. Paper read at International AAAI Conference on Weblogs and Social Media (ICWSM).

Waller, Lance A., and Carol A. Gotway. 2004. *Applied Spatial Statistics for Public Health Data*. Vol. 368. New York: John Wiley & Sons.

7

Geographic Information Systems and Disaster Recovery

CHAPTER OBJECTIVES

Upon chapter completion, readers should be able to

1. understand the different time scales at which disaster recovery operates and the implications of those time scales on the use of Geographic Information Systems (GIS) and disaster recovery,
2. discern various geographical aspects of disaster recovery and how those geographical aspects might be uniquely supported with GIS,
3. understand the concept of geocollaboration and how this theoretical idea is particularly relevant to disaster recovery and also to the use of GIS in other disaster management phases,
4. identify specific GIS techniques that can be used to support disaster recovery,
5. understand the unique role that GIS can play in recovery planning processes that involve community members, and
6. discern the overlaps between disaster recovery and disaster mitigation and how GIS and corresponding spatial data can serve both disaster recovery and mitigation needs.

INTRODUCTION

Disaster recovery is focused on the transition of the built environment, business, people, and their communities back to a state of acceptable operation after an event such as an earthquake or hurricane, which requires long-term planning and commitment to achieve recovery goals. Disaster recovery will operate at varying space and time scales subject to the nuances of the places undergoing the recovery (Stevenson et al., 2010). For example, disaster recovery can also be seen as a part of the disaster response phase in the case of short-term recovery efforts such as returning people that have been temporarily displaced to their homes. Furthermore, disaster recovery can be viewed as a disaster planning activity in terms of developing plans for recovery such as contracts for debris removal that are implemented once an actual disaster occurs or making observations of disaster zones using remote sensing technologies to measure redevelopment progress (Wagner, Myint, and Cerveny, 2012).

Furthermore, long-term disaster recovery often does not receive the media attention that other disaster phases do, such as disaster response. Thus, the use of GIS must be developed to a capacity that it can remain available and operational for the duration of a long-term disaster recovery and not simply be a novel technology that is used to help with the disaster response but then forgotten about once the immediate disaster situation stabilized. This issue is particularly pronounced at the international level when disasters strike countries that have a low existing state of GIS capacity including lack of computing infrastructure, reference datasets, skilled GIS personnel, lack of effective disaster management culture, and other context-specific issues that require external GIS support and assistance, which may eventually disappear once the initial recovery and stabilization has occurred. Thus, any GIS assistance provided for long-term disaster recovery must also include plans for the long-term sustainability and transition of GIS capacity to relevant stakeholders (Environmental Systems Research Institute, 2007).

Long-term disaster recovery (which is the focus of this chapter) makes for novel use of GIS in the overall disaster management cycle in that the process of rebuilding, redevelopment, rethinking, and planning of communities are clearly connected to GIS roots in geography, planning, and overall spatial thinking, topics which are discussed in the following section.

GEOGRAPHICAL ASPECTS OF DISASTER RECOVERY

Figure 7.1, taken from the US Federal Emergency Management Agency (FEMA) National Recovery Framework, is a helpful guide for outlining various geographical aspects of disaster recovery. The specific uses of GIS as a support mechanism for many of these geographical aspects of disaster recovery are discussed later in this chapter.

For example, in the short-term recovery stage of transition from mass care and sheltering to regular housing, GIS can be used for location-specific planning and coordination such as identifying people in specific shelters, identifying specific locations to which they can be moved, and monitoring the rebuilding and redevelopment of houses and neighborhoods. As also seen in Figure 7.1, debris and infrastructure activities are inherently spatial in nature and can rely on GIS for planning and coordination. Public health and health care are also very location-specific activities that can rely on GIS for site selection problems such as determining the best locations to place temporary health centers. Finally, it is interesting to note in Figure 7.1 that disaster mitigation activities, such as identifying risks and vulnerabilities and communicating with community members about opportunities for more resilient rebuilding, are also integrated into the disaster recovery process. Disaster mitigation is the topic of Chapter 8 and the use of GIS for activities such as risk and vulnerability assessment are discussed further in that chapter.

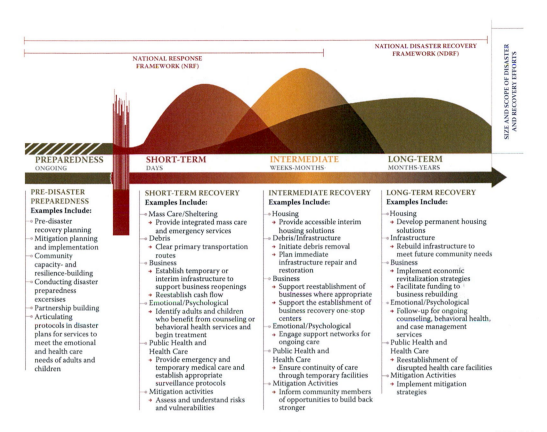

Figure 7.1 The recovery continuum. (From Federal Emergency Management Agency (FEMA). 2011. National Disaster Recovery Framework, 8. http://www.fema.gov/national-disaster-recovery-framework)

Although not specifically mentioned under the emotional/physiological category, one other geographical aspect of disaster recovery is restoring people's pride in sense of community and sense of place. Sense of place is the idea of emotions, attachments, and feelings people have for the places they live (Tuan, 1990). For example, people's sense of place can often be literally viewed during disaster recovery on informal messages spray-painted on damaged buildings, walls, or other locations to show community defiance against nature, a sense of pride that the community will return, or religious connections to a place (Figure 7.2).

USING GIS TO SUPPORT DISASTER RECOVERY TASKS

Geocollaboration

Geocollaboration is the idea of using maps, spatial representations, and map annotations to facilitate processes of collaboration that themselves are spatial in nature (Tomaszewski, 2010). Although relevant to all disaster management phases, the idea of geocollaboration

Figure 7.2 A damaged wooden fence containing the message "GOD BLESS BILOXI" (Mississippi). This picture was taken in 2006, or one year after 2005's Hurricane Katrina event where the beginning phases of long-term recovery were just starting in Biloxi. (Photo by Brian Tomaszewski.)

can play a particularly important role in disaster recovery as a means to coordinate the spatial activities of a variety of actors involved in long-term recovery (Emrich, Cutter, and Weschler, 2011).

MacEachren (2005) outlined a framework for the particular role of visualization as a support mechanism in the geocollaborative process. The MacEachren (2005) framework, applied to disaster recovery tasks, is outlined as follows:

1. *The object of collaboration:* Maps can serve as the manifestation of common ground through which collaboration and group work will occur. For example, using maps as the object to coordinate the efforts of teams working together on recovery tasks such as debris removal, deciding where to relocate businesses, or reviewing housing recovery plans (Figure 7.3).
2. *Provide support for human dialogue, information sharing, negotiation, and discussions:* Maps and visual artifacts placed on maps can be used to reformulate and re-express concepts, visualize opinions, or be used for information sharing (Figure 7.4).
3. *Support for coordinated activity:* Similar to the ideas discussed in Chapter 6 and emergency operations centers, maps, and spatial data are fundamental to supporting coordinated activities. This use of visualization to support group work is particularly (but not exclusively) important in short-term recovery following disaster response activities. For example, large-sized map displays can be used to show locations of various entities relevant to a situation (Figure 7.5).

Figure 7.3 A housing recovery being held after 2012 Hurricane Sandy. In this image, the two people are using a map as the object of collaboration to review changes in FEMA elevation flood maps. (FEMA photo by K.C. Wilsey.)

Figure 7.4 A 2010 image of a paper map of Tennessee used as a base map to show the locations of disaster recovery centers using artifacts such as colored push pins that signified the open (green), proposed (yellow), or closed (red) status of a disaster recovery shelter during major statewide storm and flooding events. This is a classic example of a low-tech but highly effective way that maps and artifacts placed on the maps, such as the push pins or hand-drawn map items, can be used for dialogue and information sharing among disaster recovery teams. (FEMA photo by David Fine.)

Rotterdam is one of the busiest ports in Europe and large-size map displays of the harbor entrance are vital for maintaining situation awareness about the position and details about ships that come and go from the harbor. Displays like this are vital for coordinating the activities of a wide variety of people involved in harbor management activities and when harbor emergencies occur.

Figure 7.5 A picture of the large map display from the port of Rotterdam in the Netherlands. (Photo from collection of Brian Tomaszewski.)

Restoring Critical Infrastructure

The restoration of critical infrastructure, such as power, water, electricity, and transportation systems, is a critical activity for both short-term and long-term recovery efforts. GIS can play an important role in critical infrastructure vulnerability planning and restoration activities through the visualization of physical proximity and distribution of critical capabilities across a region. Since infrastructural vulnerabilities are governed by virtual and physical attributes, use of GIS is particularly important as a management tool as it is a method that can marry both sets of conditions. The following is a specific GIS example using a road blockage (Figure 7.6).

Figure 7.6 shows a hypothetical critical infrastructure restoration example of conducting an analysis as to which barriers should be removed to restore optimal and efficient transportation of elderly people from shelter and health and human service locations. In this example, starting on the bottom right, the travel destination start point, which is the location of a shelter specializing in elderly people, is shown with a circled number one. A travel destination midpoint, which is the location of a pharmacy for collecting medical supplies, is shown with a circled number 2, and the endpoint, which is where the elderly people are being taken for doctor visits, is shown with a circled number 3. The ideal route for travel between these three locations is shown with the black, dashed line. However, travel along this exact route is not possible as obstacles, collected from field damage assessment teams, are blocking passage along the route. For example, there is a large debris

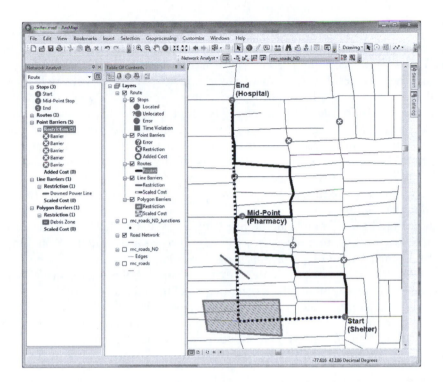

Figure 7.6 The networking analysts tool environment inside ArcMap 10.2. (Copyright © 2014 Esri, ArcGIS, ArcMap. All rights reserved. Used with permission.)

zone (represented by a hatched polygon symbol) as well as smaller debris zones causing route blockages (as represented by point symbols shown with a circled X). Both the travel route origins and destinations, in endpoints as well as point, line, and polygon restriction representations, can all be added into GIS for planning the final route. Furthermore, the underlying road network is a TIGER shapefile (see Chapter 3) that has been processed by the ArcGIS Network Analyst tool to allow for route calculations. Once the origins, destinations, and endpoints have been defined, in addition to point, line, and polygon-based route restrictions, the network analyst tool will then create a route that connects the origins and destinations while avoiding the restricted areas.

In Figure 7.6, the final route calculated is shown using a solid black line. Make note of several things about the calculated route. First, note how the calculated route connects all three of the destination points (the shelter, pharmacy, and hospital). Second, make note of how the final route avoids all the defined restriction areas. The Figure 7.6 example only shows the calculation of a route that is based on distance. However, routes can also be calculated in terms of time restrictions, such as avoiding streets with particularly high traffic volumes at certain times, such as morning or afternoon rush hour traffic. In addition to helping to restore critical infrastructure, network analyst approaches like the one shown in Figure 7.6 have a wide variety of other uses in other disaster management activities such as evacuation route planning (as discussed in Chapter 5).

Debris Cleanup

Another very common activity in disaster recovery on both short-term and long-term scales is debris cleanup. Again, GIS can serve as a powerful support technology for planning, analyzing, and modeling debris cleanup activities. For example, debris cleanup is the initial activity that signals the shift into recovery from the response phase. It is also a resource- and capability-intensive process. The right tools and equipment must be available to the individuals assigned to complete the large-scale effort. Careful planning is essential for efficient and effective debris cleanup such as (1) understanding the volume and type of debris to be removed, (2) where specific debris can be moved (for example, a general landfill versus a specialized waste facility location that specializes in hazardous materials), and how much time debris removal will take with the resources that are available (Dymon and Winter, 2012).

In addition to basic spatial understanding of where debris has accumulated based on damage assessment maps and where debris cleanup crews are located, the analytical capabilities of GIS can also be used for planning debris cleanup activities using networking algorithms similar to those shown in the previous examples of restoring critical infrastructure. Figure 7.7 is one example of the use of GIS as an analytic support device for debris cleanup activities using service area networking algorithms.

Figure 7.7 shows a hypothetical degree cleanup scenario that was created using the Network Analyst tools of ArcMap 10.2 and was also used to create the routes shown in Figure 7.6. In Figure 7.7, the black dots on the map represent debris collection facility

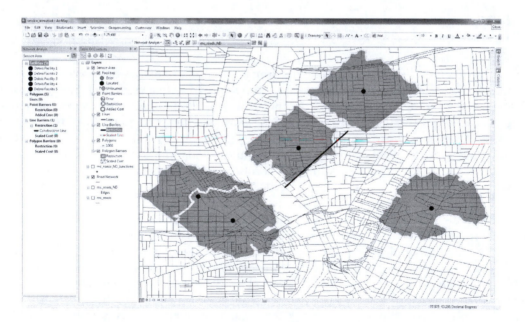

Figure 7.7 Debris cleanup service area network algorithm output results—areas 1000 meters from debris collection points. (Copyright © 2014 Esri, ArcGIS, ArcMap. All rights reserved. Used with permission.)

points, or areas where citizens have been instructed to bring debris created during a flooding episode or perhaps where local recovery volunteers have been instructed to pile debris for later collection by a large-volume debris collection truck. The dark gray polygons around each debris collection facility point represent the service area of the debris collection point. A service area, in the context of Figure 7.7, represents a set distance, represented as a polygon, from the debris collection facility point that can be traveled on the underlying transportation network. In Figure 7.7, this set distance is 1000 meters. This distance, of course, can be changed to whatever distance is required. The service area can also be defined from the facility point based on a time value opposed to a distance value. Time might be used in the case of service areas that are time dependent such as, for example, the speed at which vital medical supplies can be delivered from a hospital or temporary medical facility.

In addition to the service area polygons that represent a 1000-meter distance that can be traveled on the underlying road network from the facility points, make note of several other items in Figure 7.7. First, make note of the two service areas that overlap (bottom left of the map) and the light gray line that goes through the two service areas. This gray line represents the exact boundary where the two service areas overlap. An important characteristic of the service area–generation algorithm, showing service area overlaps can be very useful for identifying inefficiencies in service areas such as the overlapping, redundant service areas like those shown on the bottom left of the map in Figure 7.7. Additionally, the service area–generation algorithm can help to visually identify service area gaps. For example, the service area shown on the bottom right of the map Figure 7.7 has a somewhat large, open gap between the two service areas that are above and slightly to the right of the service area on the bottom right. Thus, this analysis could be used to inform decisions about where to place a new debris collection area that would also be accessible within 1000 meters but do not overlap with existing services. Also, make note of the service area located in the middle of the map in Figure 7.7. In particular, note how there is a black line running from the bottom left to the upper right of the right side of this service area. The black line represents a restriction or barrier, similar to the idea of barriers and restrictions you learned about in the restoring critical infrastructure section in this chapter and shown in Figure 7.6. Much like barriers that can restrict where routes can go, barriers can also serve to restrict or confine the definition of service areas. This can serve as a very useful feature, especially in the case of disaster recovery, for ensuring that relevant situational factors are accounted for when defining service areas. For example, the line could represent a construction zone, downed power line, or any other type of feature that citizens or volunteers helping with disaster recovery activities should avoid. Finally, like the routing algorithm discussed previously, the service area definitions can easily be modified such as moving facility locations, modifying service boundary destinations, or adding point, line, or polygon restrictions.

Recovery Planning

Recovery planning is an ideal opportunity to involve the public in broader disaster recovery activities. Once again, GIS can play an important role in facilitating recovery planning processes. Current disaster recovery policy in the United States strongly emphasizes ground level, community involvement for a wide range of stakeholders ranging from individual citizens to local businesses after a disaster has affected a community (Phillips, Neal, and Webb, 2012; Federal Emergency Management Agency [FEMA], 2011). The emphasis on local stakeholders creates ample opportunity for the broader public to engage with the products of GIS, if not GIS technology itself.

For example, maps can be used as the visual, spatial representations of ideas, arguments, and discussion points that focus on how a community rethinks and reimagines itself after a major disaster has physically, psychologically, and economically impacted the community (Rinner, 2007) (Figure 7.8).

Figure 7.8 shows a low-tech, but highly effective method for gathering inputs from community stakeholders using map-based formats. This picture, taken in 2006 in East Biloxi, Mississippi, one year after Hurricane Katrina, shows community members utilizing the second idea of the role of visual representation in geocollaboration discussed previously in this chapter—*provide support for human dialogue, information sharing, negotiation, discussions.* In particular, these community members are adding posted notes, hand-drawn annotations, and other artifacts on acetate overlays on paper maps to discuss how East Biloxi should be redeveloped in the wake of Hurricane Katrina. East Biloxi was a particularly interesting case in this regard as its built environment was almost completely destroyed by Hurricane Katrina and the community faced many redevelopment issues a such as balancing waterfront, casino development interests with that of poor and underrepresented groups that live in the interior parts of East Biloxi near the casinos (Mississippi Renewal Forum, 2005).

Figure 7.9 also shows other uses of hardcopy maps to engage the public in recovery planning activities.

Figure 7.9 was taken during a Hurricane Sandy housing recovery session with the New York City Housing Recovery Program and joined by FEMA. As seen in this figure, community members can find their house on a large map depicting a neighborhood impacted by Hurricane Sandy and can mark their house with a green (prefer to stay), orange (prefer to move), or blue (I am undecided) dot to indicate their housing situation

Figure 7.8 Using paper maps and annotations for a recovery planning session (2006) in Biloxi, Mississippi, post-Katrina. (Photo by Brian Tomaszewski.)

222

Figure 7.9 Using large maps to capture public opinion in feedback for housing recovery/restoration planning after Hurricane Sandy. (FEMA photo by K.C. Wilsey.)

preference in the wake of damage caused by Hurricane Sandy and revised flood elevation maps that were published by FEMA after Hurricane Sandy.

Thus, although GIS is often thought of in terms of a technology-based solution, it is important to remember that when GIS and the products created by GIS are used for activities such as recovery planning that involve the public, it is still very useful and valid to use simple, easy-to-understand paper-based maps on which people can draw, add Post-it® Notes, or use any other simple data collection device so the widest range of stakeholder views can be captured and incorporated into broader recovery planning and decision-making processes.

The final section discusses recovery activities that transition into mitigation activities.

TRANSITION FROM RECOVERY TO MITIGATION

As discussed in the beginning of this chapter, disaster mitigation activities can be intertwined with disaster recovery activities. The reason for this is that, as the built environment, the community, and any other aspects of a local geographical context are recovered, restored, and replaced, it is the optimal time to incorporate mitigation strategies, for example, reconstructing buildings to be more resilient to earthquakes or moving houses outside of flood zones.

One of the best examples of the use of GIS at the transition from recovery to mitigation is flood elevation maps (Figure 7.10).

Figure 7.10 is an example of a flood elevation map from the New York City area of the United States using data acquired through a mapping web service provided by the FEMA Flood Hazard Resources Map (http://fema.maps.arcgis.com/home/item.html?id=2f0a884 bfb434d76af8c15c26541a545).

Data sources such as these can provide valuable inputs to create final GIS end products such as that shown in Figure 7.10. Flood elevation maps are vital to both disaster

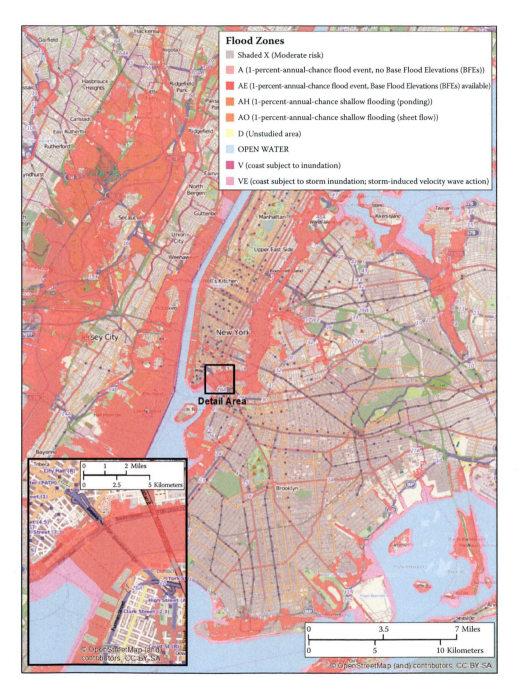

Figure 7.10 Flood elevations around the New York City region derived from an online web mapping service. (Map by Brian Tomaszewski using FEMA and OpenSteetMap data.)

recovery and mitigation because knowledge of flood levels has important ramifications for determining if flood insurance can be purchased through the National Flood Insurance Program (http://www.fema.gov/national-flood-insurance-program), application of flood plain building regulations, and general risk assessment. Development of flood risk maps can also be used for interaction with the public to communicate about potential hazards and risks where people live, thus serving as a form of mitigation if people are willing to take steps to mitigate against flood hazards.

The next chapter section is an in-depth interview with a top geospatial expert from the US federal government who shares his perspectives on GIS for disaster recovery and the other disaster cycle phases.

INTERVIEW WITH DAVID ALEXANDER: US FEDERAL GOVERNMENT GEOSPATIAL TECHNOLOGY LEADER AND EXPERT

David Alexander's (Figure 7.11) career in the geospatial information field spans more than three decades. He has practical experience at the local, state, federal, and private sector levels. Prior to his current work as a geospatial technology expert in the US federal government, David Alexander served as the Deputy Chief for the FEMA Enterprise Geospatial Services Branch, with the private sector supporting a mix of federal and state clients, and in the state of South Carolina as the Chief of Digital Cartography providing leadership to support statewide geospatial public safety and emergency management activities. He has led several national initiatives including the Department of Homeland Security (DHS) Geospatial Concept of Operations (GeoCONOPS; discussed in Chapter 4), the DHS Geospatial Information Infrastructure (GII), the Homeland Infrastructure Foundation Level Data Working Group (HIFLD), the US Department of Labor's CareerOneStop service locator system supported under grant to the state of Minnesota, and served as technical lead for

Figure 7.11　David Alexander.

the Department of Health and Human Services secretary's Operations Center Response technologies program, and administered the statewide E-911, legislative redistricting, and health and demographic mapping programs for the state of South Carolina.

Mr. Alexander has received several awards and accolades in his tenure with the US federal government for both individual and team achievements. *Federal Computer Weekly* recognized him as one of the Top 100 Federal Employees in 2013. The Armed Forces Communications and Electronics Association (AFCEA) Bethesda chapter presented Mr. Alexander an award in 2012 for Outstanding Achievement in Government-wide Initiatives—Geospatial. He has led several award-winning teams including the GeoCONOPS Project, 2012 finalist from Excellance.gov; DHS COP System, 2012 winner of Top 10 Government Systems from *Government Computer News*; DHS COP System, 2013 Winner of Distinguished System in Government from the Urban and Regional Information Systems Association (URISA); DHS GII, 2013 winner of Distinguished System in Government from URISA; and Geospatial Management Office, 2013 winner of Esri Significant Achievement Award. Mr. Alexander has also authored several publications on geospatial technology and served as a featured speaker for numerous government and organization events.

Mr. Alexander holds several advanced degrees encompassing history, geography, and business management. He is currently at the George Mason University pursuing a PhD in earth systems and geospatial information science. The following is the first of a two-part interview conducted for this book with Mr. Alexander in March 2014. In this portion of the interview, he answers questions about the role of geospatial technology for disaster management within the US federal government. The second half of this interview is presented in Chapter 9 where Mr. Alexander provides advice on getting a job in the GIS industry for disaster management with the federal government and the future of GIS for disaster management.

What types of geospatial technology, GIS-related disaster management activities does the federal government do?

We really use as many technology sources and methods we can share. So, it runs the gamut of traditional commercial products like Esri, Google Earth, Erdas Imagine, and Envi to on-premises custom government solutions for process and disseminating information across the community.

However, I think what's often undervalued is the use of geospatial technology. In this regard, I think there are two trends occurring in this discipline. One is clearly a consumerization of the capabilities. For example, just about everybody now has the capacity to fax us maps, or use a material viewing solution like Google Earth to Google Maps on their iPhone to a high-end analytic capability like Esri on their desktop. Maps are everywhere. This is a positive thing because it means that we're gaining more traction and relevancy and informing and enabling a wider variety of missions.

The other trend that's happening, which is also very useful, is that we're seeing a specialization of geospatial. It's becoming a more high-value asset to the community. We're no longer just perceived as map makers. We're truly starting to add analytic value in terms of how do we understand, interpret, and comprehend both the terrestrial landscape and the confluence of activities that may be occurring in that landscape either from a local phenomenon perspective of an incident on the ground to a global view of how we're sourcing support to that incident and whether or not the incident is cascading to affect other areas. I think those are helpful trends, although it does take some people out of their comfort zone.

Then, when you think beyond technology like I just discussed, then it is important to talk more about trade craft. For example, you have information and the fact that we see a rapid consumerization of the technology has also translated into exponential growth in the amount of geospatial information available, the fidelity of that information, and the diversity of the information. These facts offer tremendous potential both in the planning, the orchestration, and the conduct of operations. That percolates into one of my pet peeves is that geospatial and homeland security and emergency management tends to be an afterthought, not a forethought. Geospatial is starting to be included as part of the mission operations. But geospatial, in some ways, has been a niche. We haven't necessarily developed that culture of preparedness that other elements of disaster management and homeland security has. So, we don't take the old the motto from a firefighter, "train like you fight, fight like you train." It doesn't work. Geospatial doesn't usually have that. We're not necessarily good at trying to figure out what we do well. For example, identifying specifically what we did well and continue to do that, and what we didn't do well at and try to do better. Some of that, I think, is because we have been, in many cases, perceived as an afterthought. So, we're often late to the scene and our products, in many cases, are delayed. We don't have a fast time-to-market.

The other thing that I think we face is that our products have traditionally been clumsy and complicated. I think that's one of the trends we're overcoming through consumerization where we're making our product easier to use. We're getting the products faster to market. We're trying to make them more intuitive and understandable. We're trying to parse down a large volume of data into more digestible, actionable information. I think that's a healthy trend and path forward. But I do think there's still a lot of growth in terms of technology and data. I think remote sensing offers a tremendous part of the solution to the equation of situational awareness.

I think trade craft is also very important. We do not have real codified curriculum that allows us to certify a trade craft for geospatial emergency management. We've got certificates in GIS. We've got certificates in emergency management. But unlike DOD [Department of Defense] and the intel community, we don't have a real course curriculum driven towards the trade craft of GIS for emergency management. I think that's partly because we're still maturing in that area. Also, I think it's partly because we haven't really come together and championed that objective. I think this needs to start with skills and competencies. As an example of the kind of competency I think is needed is that if you want to do GIS for emergency management, disaster management, or public safety response or any of those, we know that you're going to be looking at imagery. So, you need to understand how to do imagery analysis. That's a core competency. At the same time, if you're going to do GIS in DOD or NGA [National Geospatial Intelligence Agency], you're going to do imagery analysis. Where the skill part comes in is how you apply that imagery analysis competency in the battle space is much different than

227

how you apply it in emergency management. In the battle space, you're looking for chemical discharges and you're looking for maybe ammunition signatures. You're looking for casualties or other war fighter activities that can be derived from terrestrial information. In disaster management you're looking for a totally different scenario that still can be extracted from imagery but is not the same information type. So, for example, you may be looking for debris. You may be looking for damage assessment. You may be looking for inundation. You may be looking for liquefaction. You may be looking for a whole series of other man-made or disaster environmental-related outcomes. It still requires a basic competency in imagery analysis, but you need to learn the skill set around how you apply those competencies to derive the information relevant to that domain. This example is just a starting point and an easy explanation for most people to understand, but you can transcend that example into other examples such as what other kind of vector analyses do you need to do?; what kind of vulnerability analysis do you need to do? Doing vulnerability for a disaster theater may be a little bit different in terms of the indicators and factors of doing countermeasures and analysis in a battle zone. I think that's an area where we still have a lot of work to do.

It comes back to the fact too that we need to define where the value proposition is for geospatial and whether or not geospatial, as a specialty, needs to perform that function or whether or not we just need to push the technology into the hands of the operator. This is where some of the discomfort can come in. As you start to lose span of control; you start to get nervous about your role. So, there's opportunity of risk. The opportunity is that you push your capabilities faster to the front lines, to the tactical edge. Now, the firefighters and the EMTs, the police officers, the emergency managers that aren't GIS specialists can do their job more effectively. But at the same time that's where it drives you towards specialization because now you're going to focus on not just pushing the data out in a viewable format. Additionally, you are now going to look at what interpretations can I add to the scenario that the emergency manager can act on that would make them more effective. In a way, I think we are becoming more of a back office function but also a more relevant function. We just have to really beat our time to market. Furthermore, we have to simplify our delivery. Our product can't be overcompli-cated. If it takes a rocket scientist to understand what the map is telling you, then the map is probably not that useful to begin with.

Any events, activities, situations in the last five years where you feel GIS really did demonstrate its value for disaster management at the government level?

There have been a few. Some of them may have had negative press, but I think when it comes down to it, there would be recognition that geospatial and remote sensing played an important role in terms of, if not the immediate response, but at least the recovery and postresponse. For example, Hurricane Katrina. We used remote sensing and aerial imagery to help streamline the identification of damaged structures ranging from both individual residential structures to public infrastructure. That helped us to quantify the amount of resourcing that needed to be provided back to the states as well as prioritize where debris removal and other activities needed to occur. We may have been late on arrival, but I think once we got there, we were able to apply geospatial technology effectively. As another example, I think in the 2013 Oklahoma tornados recently, the federal government was able to provision both satellite and aerial imagery almost near real time. So, we had satellite imagery flowing within hours of that event occurring. We had aerial imagery, high-resolution aerial imagery coming in within 12 hours to 24 hours. I think that helped to quantify as well as qualify the impact to the local community.

You seem to hear a lot about GIS for disaster response and immediate recovery, but do you think that GIS is equally applicable in the planning and mitigation, basically those other phases that maybe do not get as much of the media attention that happens during a big disaster but are equally as important in terms of GIS support?

I think it's crucial in planning and mitigating. I think the mitigation is the foundation of emergency management. We could reduce severity of many of the events that occur if we would recognize the vulnerabilities and risks that we're going to encounter and implement appropriate measures or put appropriate infrastructure or mitigators in place. I think GIS has proven itself in that domain. That doesn't necessarily mean it's translating to the best result and outcome, but it has proven itself highly relevant to the success of the domain. The fact that we're starting to create hazard/risk scapes, understanding where the highest probability of earthquakes are going to occur across the nation and the globe in some cases, being able to generate a risk-scape around floods is a step forward. If you start to outline all of those natural scenarios, the knowledge base starts to lend itself to the more man-made event phenomenon because you're going to use similar concepts; you just may apply and instruct them in different areas. For example, we have the emergence of infrastructure protection and cyber security. They really go hand in hand. You may have what you consider a cyber-virtual attack. I can tell you that in many cases, that attack cascades into an impact on physical infrastructure. It impacts a circuit that's sitting somewhere. It may impact an Internet depot that's somewhere. It may impact a fixed tower that stops transmitting somewhere. So, cyber, in more cases than not, has a physical instantiation, and you can apply some of those traditional constructs of risk management and mitigation to that phenomenon. Where is that infrastructure? How vulnerable is it? What are the cascading effects of that infrastructure disruption? So, those are areas that, I think, are nontraditional or at least aren't the most highly publicized where geospatial still has a tremendous and emergent impact on how we execute on those missions.

Finally, I wanted to say that I think location is everything to homeland security and emergency management. We're dealing with all hazards and all threats. Those all occur somewhere in someplace. I really don't think, just like in 911, you can live without maps and geospatial technology. But at the same time I think we also need to make sure that we're producing products that really can inform decision making. We're not just making maps for the sake of making maps. I think sometimes we get caught up on how many maps we made, not whether or not our maps made it to the front lines and actually reduced the time to save a life. For example, instead of defining people that were at risk or identifying areas that maybe have been disproportionately affected, we just get hung up on the cartography and the joy of map making and not necessarily as much on the result that we're trying to communicate.

CHAPTER SUMMARY

In this chapter you learned about GIS for disaster recovery. You were first introduced to the ideas of short-term and long-term disaster recovery and the implications these time scales have on specific use of GIS for disaster recovery. For example, short-term disaster recovery is typically intertwined with disaster response-type activities, whereas long-term disaster recovery requires sustained GIS commitments and capacity to support recovery planning and implementation activities.

Next, you were given some perspectives on the geographical aspects of disaster recovery via the various transition periods that exist during the recovery phase such as the transition from temporary housing to permanent housing and debris removal activities. Furthermore, you were given some ideas on deeper geographical aspects of disaster recovery in terms of restoring the community's sense of place after a disaster, a topic that is ripe for further exploration and research on the use of GIS to help rebuild lost sense of community and place after a disaster.

The chapter then presented specific examples of the use of GIS to support disaster recovery. The first was a theoretical discussion of the topic of geocollaboration, or the use of GIS and related spatial and visual artifacts to support group work. In particular, you learned about MacEachren's (2005) three uses of visualization to support geocollaboration with specific examples drawn from disaster recovery.

Next, you learned how GIS can be used for restoring critical infrastructure using an example of the networking algorithm in the Esri Networking Analyst tool. This was followed by a similar discussion on the use of networking algorithms for defining service areas for debris cleanup. Both of these examples illustrate the analytical power of GIS for disaster recovery decision making and scenario modeling. The last example on the use of GIS to support disaster recovery involved ideas on how GIS and GIS-derived products such as paper maps can be used to support community recovery planning activities. These were important ideas to consider in terms of the nontechnical uses of GIS for the broader public participation in disaster recovery activities such as using paper maps that people can draw on to provide feedback to decision makers and other stakeholders.

The chapter then showed examples of how, specifically, the transition from recovery to mitigation can occur. In particular, the example of flood elevation maps was used to show the duality between recovery activities and mitigation activities such as rebuilding houses damaged during a flood to ensure that the house is more resilient to future floods. The chapter concluded with an interview of a US federal government geospatial technology expert and leader. The next chapter expands on ideas first introduced in this chapter and specifically discusses the use of Geographic Information Systems for disaster mitigation.

DISCUSSION QUESTIONS AND ACTIVITIES

1. Although the US National Disaster Recovery Framework does not specifically discuss GIS, what are some ideas for the use of GIS for disaster recovery that you can find in the National Disaster Recovery Framework? See http://www.fema.

gov/media-library/assets/documents/24647?fromSearch=fromsearch&id=5124 to find the National Disaster Recovery Framework. How do those ideas tie in with the US National Response Plan that you first learned about in Chapter 6?
2. With the advent of modern collaboration tools such as Facebook and Twitter, how would you utilize the ideas of geocollaboration to take advantage of modern social networking and collaboration tools with a spatial focus for disaster recovery?
3. How might you design a technological solution utilizing GIS to support the activities outlined in Figure 7.4, which were based on a paper map and push pins?
4. How might the ideas of crowd sourcing and crisis mapping like those discussed in Chapter 6 be used for community recovery planning activities?

RESOURCES NOTES

For more discussion on GIS for critical infrastructure protection, see CRC Press for a description of the book *GIS for Critical Infrastructure Protection*, http://www.crc-press.com/product/isbn/9781466599345.

For more information on networking algorithms, see the Esri Network Analyst Tool Reference at http://www.esri.com/software/arcgis/extensions/networkanalyst.

QGIS Network Analysis Library (open-source alternative): http://www.qgis.org/en/docs/pyqgis_developer_cookbook/network_analysis.html.

To learn more about FEMA Region II flood mapping activities, see http://www.region2coastal.com/home.

For more information on the data categories shown in Figure 7.10, see https://msc.fema.gov/webapp/wcs/stores/servlet/info?storeId=10001&catalogId=10001&langId=-1&content=floodZones&title=FEMA%2520Flood%2520Zone%2520Designations.

REFERENCES

Dymon, Ute J., and Nancy L. Winter. 2012. *Supporting Emergency Recovery Operations (Debris Management) Using GIS*, Federal Emergency Management Agency (FEMA), http://training.fema.gov/EMIWeb/edu/docs/hrm/Session%2012%20-%20Supporting%20Emergency%20Recovery%20Operations.ppt (accessed March 16, 2014).

Emrich, Christopher T., Susan L. Cutter, and Paul J. Weschler. 2011. "GIS and emergency management." In *The SAGE Handbook of GIS and Society*. London: Sage, 321–343.

Environmental Systems Research Institute. 2007. GIS for Disaster Recovery. Environmental Systems Research Institute, http://www.esri.com/library/brochures/pdfs/gis-for-disaster-recovery.pdf.

Federal Emergency Management Agency (FEMA). 2011. National Disaster Recovery Framework, http://www.fema.gov/national-disaster-recovery-framework.

MacEachren, Alan M. 2005. "Moving geovisualization toward support for group work." In *Exploring Geovisualization*, edited by J. Dykes, A. MacEachren, and M. J. Kraak. New York: Elsevier.

Mississippi Renewal Forum. 2005. Reconstruction Plan for Biloxi, Mississippi, http://mississippirenewal.com/documents/Rep_Biloxi.pdf.

Phillips, Brenda D. David M. Neal, and Gary R. Webb. 2012. *Introduction to Emergency Management.* Boca Raton, FL: CRC Press.

Rinner, Claus 2007. "A geographic visualization approach to multi-criteria evaluation of urban quality of life." *International Journal of Geographical Information Science* 21 (8):907–920.

Stevenson, Joanne R., Christopher T. Emrich, Jerry T. Mitchell, and Susan L. Cutter. 2010. "Using building permits to monitor disaster recovery: A spatio-temporal case study of coastal Mississippi following Hurricane Katrina." *Cartography and Geographic Information Science* 37 (1):57–68.

Tomaszewski, Brian. 2010. "Gecollcollaboration." *Encyclopedia of Geography,* http://www.sage-ereference.com/geography/Article_n472.html.

Tuan, YF. 1990. *Topophilia: A Study of Environmental Perception, Attitudes, and Values.* New York: Columbia University Press.

Wagner, Melissa A., Soe W. Myint, and Randall. S. Cerveny. 2012. "Geospatial assessment of recovery rates following a tornado disaster." *IEEE Transactions on Geoscience and Remote Sensing* 50 (11):4313–4322.

8

Geographic Information Systems and Disaster Mitigation

CHAPTER OBJECTIVES

Upon chapter completion, readers should be able to

1. appreciate the particularly interdisciplinary nature of disaster mitigation and Geographic Information Systems (GIS);
2. have basic understanding of the concepts of vulnerability and resilience and how they relate to GIS;
3. understand the basics of disaster mitigation policy and how GIS is relevant to that policy;
4. be familiar with international organizations focused on disaster mitigation;
5. have a basic understanding of different social and physical spatial variables that can be used in GIS to model risk, vulnerability, and resilience;
6. understand basic approaches for developing vulnerability, risk, and resilience indexes using different types of spatial variables.

INTRODUCTION

As discussed at the end of Chapter 7, mitigation activities are often interwoven into disaster recovery activities as the recovery process is generally an opportune time for implementing disaster mitigation measures. Disaster mitigation has been defined as "[t]he capabilities necessary to reduce loss of life and property by lessening the impact of disasters" (United States Department of Homeland Security 2013, 1). GIS can play a particularly important role in disaster mitigation activities through the modeling of hazard and risk scenarios to identify potential physical, virtual, and social vulnerabilities, that can ideally be mitigated or reduced through increased resilience efforts. As an example using earthquake risks, GIS data layers can be created that inventory housing characteristics such as building material and structural types in relation to the location of earthquake fault lines

233

and landslide risks to determine how vulnerable the built environment is to a potential earthquakes (Kemp et al., 2008; Environmental Systems Research Institute, 2007). Making such determinations can then be used to inform decision making as to which buildings might require additional reinforcements to withstand earthquakes or will require higher insurance to recover losses in the event of an earthquake.

Like all disaster management cycle phases, disaster mitigation is no different in that the use of GIS for disaster mitigation will be sensitive to the nuances and idiosyncrasies of the underlying people, places, culture, history, and overall geographic context being represented and analyzed in GIS. Furthermore, disaster mitigation for GIS often requires interdisciplinary connections across multiple areas such as earth science, sociology, and environmental science. These interdisciplinary connections between GIS and other disciplines are very important in that that the use of GIS for disaster mitigation activities must be guided by clear understanding of the underlying scientific principles and processes inherent in a wide range of natural and man-made hazards, incident types, and underlying vulnerabilities. For example, the community that routinely faces flood hazards will need to have the perspectives of a flood hydrologist to effectively use spatially oriented models that examine flood dynamics such as flood frequency and how these physical processes can have potential effects on the human environment; a community that is near a nuclear power plant will require the perspectives of a nuclear engineer to understand the potential impacts of radioactive clouds in the event of a nuclear meltdown. Interdisciplinary connections between GIS and other disciplines is, in fact, a topic echoed several times by several of the GIS disaster management practitioners who were interviewed for this book as you will see in Chapter 9 when their advice for getting a job in the GIS for disaster management field is presented. If you are (or will become someday) the *GIS person* at some given organization, meaning that your job is primarily focused on operating GIS software and working with spatial data, it will be very important to learn to *speak the language* of people from other disciplines you are working with so as to be effective at developing effective GIS models and end products such as maps that properly convey the conventions, norms, and scientific principles of the other discipline. For example, if you are working with earthquake scientists, being sure that you understand the concept of the Richter scale when developing maps of earthquake hazards and using the proper visual variables such as shape, size, and color to represent the Richter scale. The following sections further discuss two closely related disaster mitigation concepts that are particularly spatial in nature—vulnerability and resilience.

VULNERABILITY

Vulnerability is generally considered to include factors that make a community or system susceptible to the effects of a hazard (United Nations Office for Disaster Risk Reduction [UNISDR], 2007b). In hazards and disaster research, the concept of vulnerability emerged out of the social sciences in response to the hazard-centric perspective of disaster risk (Schneiderbauer and Ehrlich, 2004 cited in Birkmann, 2006). For example, from the tradition of civil defense that came out of World War II, disasters were thought of as isolated events that caused a disruption in the human condition and humans were passive victims (Phillips et al., 2010).

In the 1980s, this mode of thinking began to change with disaster consequences being viewed as deriving from impacts on complex social conditions (Phillips et al., 2010). For example, Hurricane Katrina was a severe event not only because of its physical impacts, but also because of its impacts on vulnerable people such as the poor and existing geographical conditions of poverty. Social vulnerability has become an important factor in rethinking disaster management in terms of shifting focus to risk management and mitigation and away from preparedness and response/relief as people who are socially vulnerable also lack the access to ways to protect themselves from impending crisis. (Federal Emergency Management Agency (FEMA), 2010). The Geographical Sciences have seen a variety of work on the concept of social vulnerability from a hazards/disasters perspective (as opposed to vulnerability and resiliency of built environments to natural hazards). For example, Cutter et al. (2003) presents methods for modeling and quantifying social vulnerability through development of a social vulnerability index (SoVI) using county-level US census data. Ebert et al. (2009) examined physical proxy variables such as building materials observed via remote sensing and GIS for understanding social vulnerability in urban environments.

Vulnerability has also been a theoretical construct in economic science. Spatially, the concept has been used, at the household level, to make inferences in regard to future consumption patterns based on factors such as future income expectations for poverty alleviation (Chaudhuri, Jalan, and Suryahadi, 2002). The United Kingdom–based Department for International Development (DFID) explicitly defines a *vulnerability context* as a conceptual device for framing the external environment in which people live and how external forces that operate at varying space–time scales such as trends (long term), shocks (sudden onset), and seasonal (cyclical/recurring) potentially affect livelihoods (Department for International Development [DFID],1999). Livelihoods, as per the DFID framework, are based on the notion of interrelated *capitals* that provide a "soft" (i.e., nonquantitative) index for understanding social vulnerability. Many of these capitals naturally lend themselves to mapping and analysis with GIS, for example *human capital*, or knowledge, health, and skills needed for working; *natural capital*, or natural resources related to livelihoods such as water, land, and biodiversity; *financial capital* such as cash and other financial resources; *social capital* or social relationships and group memberships that people may draw upon for finding livelihoods; and *physical capital* or physical assets such as roads, clean water, and shelter that provide an infrastructure to support livelihoods (Frankenberger et al., 2002).

RESILIENCE

Often considered (metaphorically) to be on the opposite side of a coin from vulnerability, *resilience* is considered to be the ability of a system or community to withstand the impacts of an event and recover to an acceptable or existing or even an improved state in comparison to what was available before an event (United Nations Office for Disaster Risk Reduction [UNISDR], 2007a). In recent years, the concept of resilience has gained more emphasis as the concept of vulnerability has been seen to imply that people are passive victims. The use of GIS as an information management device to inventory, analyze, visually represent, and ultimately understand and manage risks as a means to improve resilience continues to grow (Fung, 2012). Cutter et al. (2013) outlines a comprehensive agenda for building disaster resilience in the United States that emphasizes understanding, managing, and reducing risk, developing resilience metrics, local resilience capacity building, and changes in resilience policy across all levels of government. Of particular interest to GIS, the report specifically outlines the role of GIS for risk reduction. Although not directly mentioned, GIS is also relevant to recommendation 3 made in the report: "A national resource of disaster-related data should be established that documents injuries, loss of life, property loss, and impacts on economic activity" (Cutter et al., 2013, 8) as the data management and analysis capabilities of GIS (as was discussed in Chapter 3) are directly relevant to the geographical aspects of developing such resources.

DISASTER MITIGATION POLICY AND INTERNATIONAL PERSPECTIVES ON GIS

The following sections briefly outline some relevant disaster mitigation policies and international perspectives important to understanding the role of GIS as a disaster mitigation support mechanism.

The United States National Mitigation Framework

The United States Department of Homeland Security published the National Mitigation Framework in 2013 to establish "a common platform and forum for coordinating and addressing how the Nation manages risk through mitigation capabilities. It describes mitigation roles across the whole community" (United States Department of Homeland Security, 2013, 1). Of particular interest to the role of GIS for disaster management activities are the seven core capabilities required of groups, organizations, and communities involved in disaster mitigation. These capabilities, quoted below from the National Mitigation Framework (United States Department of Homeland Security, 2013, 15–25), are:

- *Threats and Hazard Identification* – Identify the threats and hazards that occur in the geographic area, determine the frequency and magnitude, and incorporate this into the analysis and planning processes so as to clearly understand the needs of a community or entity.
- *Risk and Disaster Resilience Assessment* – Assess risk and disaster resilience so that decision makers, responders, and community members can take informed action to reduce their entity's risk and increase their resilience.
- *Planning* – Conduct a systematic process, engaging the whole community as appropriate, in the development of executable strategic, operational, and/or community-based approaches to meet defined objectives.
- *Community Resilience* – Lead the integrated effort to recognize, understand, communicate, plan, and address risks so that the community can develop a set of actions to accomplish mitigation and improve resilience.
- *Public Information and Warning* – Deliver coordinated, prompt, reliable, and actionable information to the whole community through the use of clear, consistent, accessible, and culturally and linguistically appropriate methods to effectively relay information regarding any threat or hazard and, as appropriate, the actions being taken and the assistance being made available.

- *Long-Term Vulnerability Reduction* – Build and sustain resilient systems, communities, and critical infrastructure and key resources lifelines to reduce their vulnerability to natural, technological, and human-caused incidents by lessening the likelihood, severity, and duration of the adverse consequences related to the incident.
- *Operational Coordination* – Establish and maintain a unified and coordinated operation structure and process that appropriately integrates all critical stakeholders and supports the execution of core capabilities.

GIS is specifically mentioned within the critical task analysis for Risk and Disaster Resilience Assessment: "Develop analysis tools to provide information more quickly to those who need it and make use of tools and technologies, such as Geographic Information Systems (GIS)" (United States Department of Homeland Security, 2013, 17). These capabilities are an important framework for ideas on how GIS can serve a multitude of mitigation activities besides assessment (see "Discussion Questions and Activities" section of this chapter).

International Perspectives on Disaster Mitigation: UNISDR

A prime example of an international organization focused on disaster mitigation is the United Nations Office for Disaster Risk Reduction (UNISDR; http://www.unisdr.org/). UNISDR is broadly mandated "to serve as the focal point in the United Nations system for the coordination of disaster risk reduction and to ensure synergies among disaster risk reduction activities" (United Nations Office for Disaster Risk Reduction, n.d.). For example, UNISDR facilitates disaster risk reduction coordination such as the Global Platform for Disaster Risk Reduction in the Hyogo Framework for Action. They also campaign for activities such as making cities resilient, safer schools and hospitals, and outreach events such as International Day for Disaster Risk Reduction. They are also involved in climate change adaptation advocacy campaigns, disaster risk reduction education, gender issues, and sustainable development practice. UNISDR is an important source of information on disaster risk topics such as maintaining a list of disaster risk reduction terminology, disaster statistics, scientific publications, and other activities of the United Nations related to disaster risk reduction. Although GIS is not an explicit activity of UNISDR, their role as an international disaster risk reduction coordinating mechanism makes them important for finding out information about the use of GIS for disaster risk reduction activities around the world, particularly in developing countries (see Fung, 2012 as an example).

GIS TECHNIQUES FOR DISASTER MITIGATION

As you have learned so far in this chapter, a specific role for the use of GIS and disaster mitigation activities is that of risk and vulnerability assessment. The following chapter sections discuss specific GIS techniques for risk and vulnerability assessment. Note that the GIS techniques provided are not specific to any particular hazard, risk, incident, or vulnerability. You are encouraged to adapt these ideas to specific geographical contexts and hazards and risks of interest and relevance to you.

Spatial Indexing and Modeling of Risk and Vulnerability

A very common technique for risk and vulnerability assessment is developing a spatial index of a risk or vulnerability level. The term *spatial index* means a numerical or qualitative value that is assigned to a preexisting spatial unit or geographical region, for example, using a numerical scale of 1 to 10, with 1 being the lowest and 10 being the highest, and assigning a number from this range that represents a vulnerability level (as determined by different variables and calculations) to a preexisting spatial unit such as state, county, or city or a geographic area. The spatial units or areas and their corresponding vulnerability spatial indexes can then be visually represented using the cartographic techniques discussed in Chapter 2 such as color hue to indicate a risk or vulnerability level in order to support vulnerability assessment. The following sections further discuss these ideas.

Social Variables

A wide variety of social variables can be incorporated into the development of risk and vulnerability spatial indexes. Social variables are important as the impacts of disasters on the social fabric of society is often where the greatest damage can occur (Mileti, 1999). For example, Hurricane Katrina in 2005 demonstrated the devastating effects that a hurricane can have on poor and elderly people (Phillips et al., 2010). A nonexhaustive list of representative, social variables used in the development of risk and vulnerability spatial indexes for a variety of natural hazards, based on Cutter, Boruff, and Shirley (2003, 246–249) include:

1. Age (elderly, children)—older people have difficulty with evacuating during a disaster; children may have a general lack of resilience
2. Gender—women can be more vulnerable due to family care responsibilities and potential lower wage earning
3. Rural/urban—rural people may be most vulnerable due to general lower wage opportunities; urban people maybe more vulnerable to complications from evacuating out of urban settings
4. Education level—lower education levels may make people more vulnerable due to the inability to understand disaster warning or recovery information.

Data on social variables can be acquired from a variety of sources. In the United States, a very common source of geographically oriented data related to the aforementioned and other types of social variables can be acquired from the US Census Bureau American FactFinder website (Figure 8.1).

Figure 8.1 shows a search on the American FactFinder website for older populations from all census tracts within Monroe County, New York (as seen in the "Your Selections" box at the top left of Figure 8.1). As can be seen in the middle of Figure 8.1, this particular search returned numerous indicators related to the general search topic. In terms of use with GIS, data can then be downloaded from the American FactFinder website in spreadsheet formats that can be combined with Census reference datasets for use in various GIS tools (see the Resources section of this chapter for further software technical guidance on doing table joins with either Esri or QGIS technology). If you are not from the United States, I encourage you to look at your country's national census department for an equivalent

238

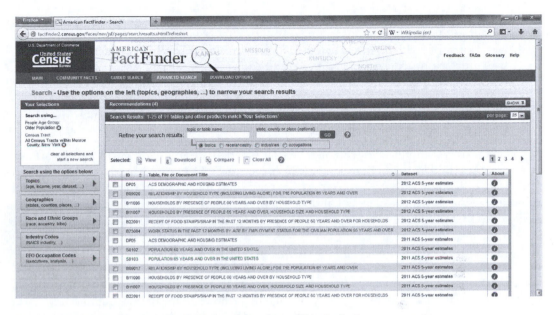

Figure 8.1 The American FactFinder website: http://factfinder2.census.gov/.

type of service. The availability of data will vary from country to country and you may even be faced with the challenge of having to extract data out of PDF-based reports if the data is not contained in an easy-to-download format like that of the American FactFinder website. So be prepared to do some digging and data engineering if you are able to find something that is of potential use.

For data at the regional scale, you are encouraged to contact your local state, county, or town governments to see if they can provide you with relevant data. For international data, a good source to use for a wide range of country-level indicators is the World Bank Data website that was first mentioned in Chapter 3 (http://data.worldbank.org/country).

Physical Variables

A wide range of physical variables can also be used to develop risk and vulnerability spatial indexes. Note that physical variables are often more clearly tied to specific hazard types, and when developing spatial indexes, it is important to consider the specific hazard type for which the spatial index will be developed. A nonexhaustive list of representative physical variables used in the development of risk and vulnerability spatial indexes for a variety of natural hazards, based on Kappes et al. (2012), Papathoma-Köhle et al. (2011), and Douglas (2007) include

1. Building material—materials from which buildings are constructed are an important variable to consider in terms of how well the building will withstand impacts or events such as earthquakes.
2. Slope percentage (%)—slope% is a very commonly used variable for determining the likelihood of landslides.

3. Proximity to flood zones—flood zone delineation is a very common measure of vulnerability and is often used for setting insurance rates.
4. Locations of critical infrastructures—critical infrastructures, such as major transportation arteries, hospitals, communication systems, power, and water, are important to the resiliency of a community in the event of a disaster and therefore often need to be accounted for when developing vulnerability models in the event that one of these or other types of critical infrastructures are impacted as a result of a disaster.

Data on physical variables can also be downloaded from a wide variety of locations. For example, the United States Geological Survey (USGS) Earth Resources Observation and Science Center (EROS; http://eros.usgs.gov/) provides access to many global-level remote sensing products such as Landsat satellite imagery, digital elevation models, and land cover that can be used as physical variable data inputs in GIS (Figure 8.2).

Using GIS to Develop Spatial Indexes of Vulnerability and Risk
One approach to developing spatial indexes for vulnerability and risk with GIS is based on a *site selection* technique. This is the idea of answering a question or testing a hypothesis by overlaying and comparing a variety of spatial variables that can be weighted and combined to select final candidates to create a final index score or match site selection criteria. This general concept is illustrated in Figure 8.3.

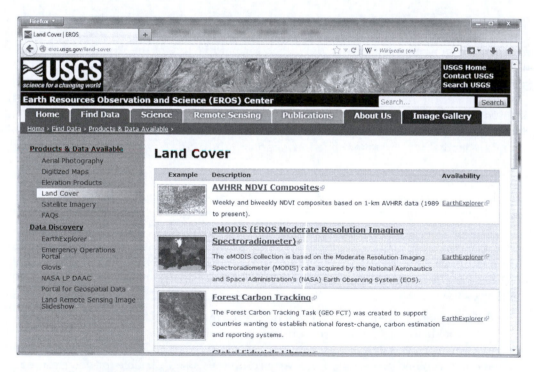

Figure 8.2   Land cover datasets available through the USGS EROS center.

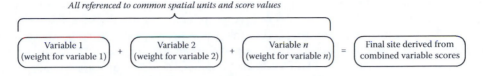

Figure 8.3   A generic GIS site selection framework.

Both raster and vector datasets can be used for site selection problems or developing spatial indexes within existing spatial units such as census tracts. All spatial variables used must be in a common spatial unit, such as the same raster grid cell size, and must use common score values. Using common score values is the idea that every variable uses a common measurement. For example, if you are combining slope values, which typically are a percentage, with distance to roads in meters, these values on their own cannot be compared as they are not in the same number system and must be normalized to a common measurement system such as a 1–10 scale. Furthermore, scores associated with a particular variable can be given a weight to indicate the importance of the variable, much like the idea of a weighted average.

The following example, based on Environmental Systems Research Institute (2014), outlines a simple, yet realistic example of developing a spatial index for risk and vulnerability assessments. This example is designed to get you thinking about what GIS can do for disaster mitigation applications and you are encouraged to take these foundational ideas in new directions based on your own research and geographic context-sensitive needs.

The vulnerability assessment model will use two social variables:

1. elderly population (as represented by census tract boundaries), with 40% influence (or weight as per Figure 8.3);
2. female population (as represented by census tract boundaries), with 40% influence (or weight as per Figure 8.3)

and one physical variable: proximity to hospitals (as represented by place name points obtained from the USGS), with 20% influence (or weight as per Figure 8.3).

The procedure for developing the model is as follows. First, each of the three variable datasets are converted from vector format to raster format. This will allow the three variables to be combined and analyzed using the Weighted Overlay Tool of Esri's ArcMap Software. The Weighted Overlay Tool allows raster datasets to be combined together and weighted based on a given layer's importance (see Arnold et al., 2012, as an example of using the Weighted Overlay Tool). Second, their respective values will be reclassified to a common measurement scale between 1 and 10 so that the vulnerability factors can be compared equally. For example, the number of elderly people, females, and proximity to hospitals will, respectively, have different numerical values and these numerical values must be reclassified to a common measurement scale in order to derive a meaningful index score. Third, the reclassified datasets will be added to the Weighted Overlay Tool and their percentages of influence will be applied for deriving the final index score. For example, if a cell in the female population raster has a reclassified value of 2 and the female raster has a 40% influence in overall score, the calculated cell value will be based on 2 * 0.4 or 0.8; if the corresponding cell in the elderly population raster has a reclassified value of 3 and the elderly raster has a 40% influence in overall score, the calculated cell value will be based on 3 * 0.4 or 1.2; and if the corresponding cell in the proximity to hospitals raster has a reclassified value of 5 and the proximity to hospitals raster has a 20% influence in overall score, the calculated cell value will be based on 6 * 0.2 or 1.2. Then, the final value for the output raster cell will be based on the sum of values from the three rasters : 0.8 (female) + 1.2 (elderly) + 1.2 (hospital) = 3.2, but since the weighted overall tool creates integer values, the final value of the cell in the output raster created by the Weighted Overlay Tool in this case would be rounded down to 3.

Figures 8.4 through 8.8 graphically outline each of the aforementioned steps in developing the model.

1. The female and elderly population vector datasets are converted to raster datasets (Figure 8.4). This allows these layers to be used in the Weighted Overlay Tool, which makes cell-by-cell comparisons.

2. A Euclidean distance function is run on the hospital data layer to determine distances to the closet hospitals (Figure 8.5).

3. The three datasets are reclassified to common measurement scales. At this point in the procedure, each dataset has completely different numerical values. For example, cell values of the elderly and female population datasets use counts based on their census tract value when they were converted to raster and the hospital Euclidean distances are measured in meters. These data need to be put into a common measurement scale such as a 1–10 scale in order to develop a meaningful index. For this example, data will be reclassified using the following logic:

    • The higher the counts of elderly and female populations, the more vulnerability exists.

    • The closer the proximity to a hospital, the less vulnerability.

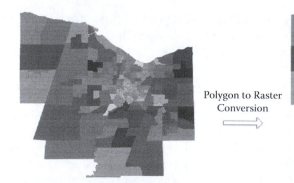

Polygon to Raster Conversion

Vector - Female Census Tracts   Raster - Female Census Tracts

Figure 8.4 Example of converting a vector dataset to a raster dataset using a polygon-to-raster conversion tool. These datasets are both symbolized using a five-class, equal interval classification scheme. (Map by Brian Tomaszewski.)

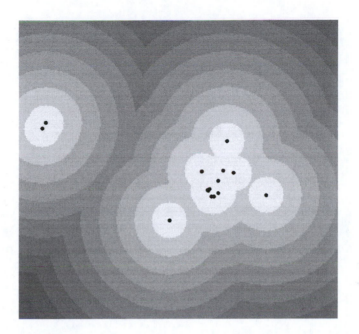

Figure 8.5 Euclidean distance calculation output for hospital locations represented as black dots. Euclidean distance is an effective raster-based analytical technique for determining distances between features.

The Reclassify tool is used to implement this logic (Figure 8.6):

4. After each of the datasets has been reclassified to a common measurement scale that fits the vulnerability model scoring logic, the reclassified datasets are then added to the Weighted Overlay Tool and the percentage of influence of each layer (40% female, 40% elderly, 20% hospital) is applied (Figure 8.7).

The final spatial index is then calculated based on values in each of the cells being multiplied by their respective percent of influence and then added to the values from the cells in the other layers. The output raster created by the Weighted Overlay Tool can then be cartographically represented using the techniques discussed in Chapter 2, such as varying color lightness, that represent different vulnerability levels (Figure 8.8).

As can be seen in Figure 8.8, the original census tract boundaries have been overlaid on top of the final output raster layer created by the Weighted Overlay Tool to visually define vulnerability by census tract. This final output layer has been stylized using an equal intervals classification, and easy-to-understand category labels such as very high, high, moderate, low, and very low have been used in the legend so that the general public, for example, will be able to understand what the map shows. Make note that in the center of Figure 8.8, which is an urban center, vulnerability is quite low as shown by the white to light gray areas, meaning that many hospitals are in this area and the female and elderly populations are low. On the outer edges of this urban area, vulnerability levels are much higher due to the opposite effects. Also make note of a spatial outlier shown on the left side of Figure 8.8 (a white area), which is the location of a rural hospital that decreases the vulnerability in

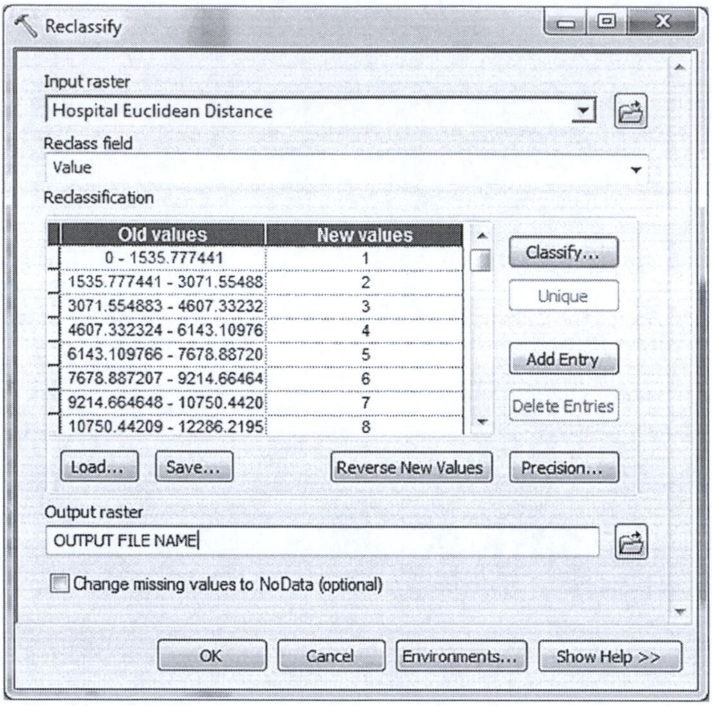

Figure 8.6   The reclassify tool. Note how the column shown on the middle left with the heading Old Values shows data class breaks from the hospital Euclidean distance that was shown in Figure 8.5. The column shown in the middle right indicates what the new values will be after the reclassify tool has run and created a new output raster. (Copyright © 2014 Esri, ArcGIS, ArcMap. All rights reserved. Used with permission.)

the general proximity of the hospital, but the overall vulnerability surrounding this hospital is somewhat high. An important point to keep in mind with using this type of approach is that the census tracts are an aggregation of data. High counts of female and/or elderly people will not be exactly located at specific raster cell locations due to the aggregation. Thus, when developing a vulnerability index like the one just demonstrated, it is important to validate the model through field surveys, citizen questionnaires, or other means of "ground truthing" the model results to further refine and calibrate model variables and their weights in order to build confidence in model results (Maguire, Batty, and Goodchild, 2005).

CHAPTER SUMMARY

In this chapter you learned about GIS for disaster mitigation. You first learned that disaster mitigation encompasses several important ideas, such as risk, vulnerability, and resilience, all of which are intertwined within the subtleties of geographic contexts. Furthermore, you learned that disaster mitigation is perhaps the most interdisciplinary

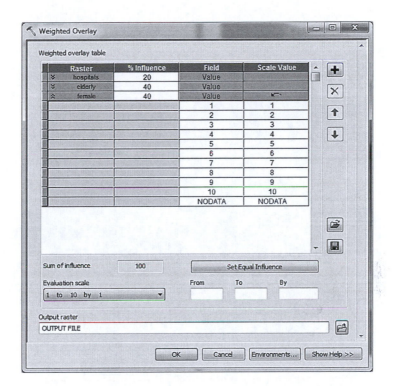

Figure 8.7 Combining and weighting the three variable layers in the Weighted Overlay tool. (Copyright © 2014 Esri, ArcGIS, ArcMap. All rights reserved. Used with permission.)

aspect of the overall disaster management cycle in that natural hazards, manmade hazards, underlying risks and vulnerabilities, and resilience require the perspectives of many different disciplines, and GIS is often the unifying platform for combining these perspectives. If your role is GIS specialist, it will be important to operate and think in an interdisciplinary manner and your own underlying spatial thinking as a GIS specialist can be brought to bear on other disciplines and potentially provide new perspectives and insights.

You then learned about disaster mitigation policy perspectives on GIS, such as the US National Mitigation Framework, which defines seven core capabilities of communities to be effective at reducing risk and building community resilience. You also learned a little bit about the United Nations Offer for Disaster Risk Reduction, which is an internationally focused organization with a particular mandate on disaster risk reduction.

You were then shown some GIS techniques for disaster mitigation. In particular, you learned about different variables that can be used to model risk and vulnerability. For example, social variables such as population characteristics (elderly, women, children) and geographical characteristics variables such as rural and urban populations. You also learned about physical variables that can be used to model risk such as slope, which is relevant to landslide hazards, proximity to flood zones, and locations of critical, lifesaving infrastructure such as hospitals. You were also shown spots where you might find specific

Final Vulnerability Index

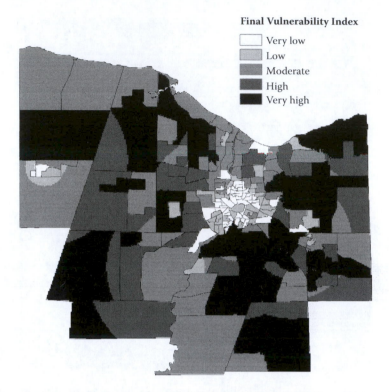

Figure 8.8   Final vulnerability index map. (Map by Brian Tomoszewski.)

data for use in GIS to incorporating these kinds of variables into risk and vulnerability models such as the American FactFinder website for US census data and the United States Geological Survey Earth Resources Observation and Science Center (USGS-EROS) for global datasets such as land use, land cover, elevation, and satellite imagery. It is important to note, that the variables listed in this chapter are by no means an exhaustive list of all possible variables that can be used for risk, vulnerability, and resilience modeling. The variables that were listed are simply meant to get you started thinking spatially about the kinds of variables used in GIS-based disaster mitigation modeling. You encouraged to read more and do your own research on finding variables that are relevant to your specific geographic context or interests.

Next, you were shown one example of how to develop the spatial index of vulnerability using the variable examples previously discussed. This is a very foundational idea that is used in a wide variety of GIS modeling application, and you are strongly encouraged to learn how to repeat these general techniques for other types of site selection and modeling problems that can be solved with GIS. Specifically, you were shown how to use US census tracts along with hospital locations to develop a simple vulnerability model that considered the interactions between female and elderly populations in relation to distance to hospitals to develop a vulnerability score that can be used to inform disaster mitigation activities such as where to increase medical attention during a disaster or making sure that special-needs populations like the elderly are accounted for in community disaster planning activities.

This chapter concludes the series of chapters that are focused on a specific disaster management cycle phases. The final book chapter takes a forward-looking perspective on where the field of GIS for disaster management is heading, advice on staying current in the GIS for disaster management field, special GIS for disaster management topics, and extensive advice on developing a career and finding a job in the GIS for disaster management field from the perspective of the many working practitioners who provided interviews for this book and who you met in previous chapters.

DISCUSSION QUESTIONS AND ACTIVITIES

1. Review the seven core capabilities of mitigation from the National Mitigation Framework presented in the disaster mitigation policy section of this chapter. What are some additional, specific uses of GIS you could foresee for supporting these capabilities in addition to the analysis capability specifically mentioned in the National Mitigation Framework? Think creatively and spatially about how GIS can be used and think back to many of the ideas you have learned about previously in this book such as using maps for public communication, geocollaboration, and field data collection and mobile GIS.

2. As you can probably imagine, developing indexes of vulnerability, risk, and other factors relevant to disaster mitigation is a massive area for scientific inquiry and application of a wide variety of GIS analytic tools that can operate on an equally wide number of underlying GIS data sources. Using the ideas for developing vulnerability indexes, along with the GIS techniques that were provided in this chapter along with your own research (use the vulnerability literature cited in this chapter as a starting point), develop a vulnerability, risk, and/or reliance model for an area of interest to you such as your hometown, county, state, or even country. What variables should you include? How would you weight them and why? If possible, try and work with other people with a wide variety of perspectives from other disciplines such as the social sciences, physical sciences, and computational sciences to incorporate interdisciplinary perspectives on the GIS model you or your team develops.

3. There are endless ways in which risk and vulnerability can be communicated in map-based formats. Do some Internet searching and find examples of either interactive mapping web sites, or published map products that are used to communicate risk and vulnerability. Look back to earlier chapters in this book for ideas from organizations such as UNISDR, UN ReliefWeb, the FEMA GeoPortal, and the United States Geological Survey. When looking at these maps, what specific GIS technologies or range of technologies might you use to create them? For example, if you had a job as the disaster management GIS person for a small town, how might you help citizens in your town create vulnerability maps of their community using free and open-source tools like those discussed in Chapter 3?

4. An idea that has been mentioned before in this book is that GIS, for all of its benefits, also has limitations in that it reduces reality to computer-based representations that often cannot convey or represent the subtleties and idiosyncrasies of geographic context. In this regard, how might GIS be used to incorporate local knowledge such as knowledge of people's communities, social bonds with one another, local environmental knowledge, and other factors not easily conveyed in computer-based representation in order to gain deeper perspectives about disaster risks that can be used to inform disaster mitigation activities (see Kemp 2008 and Tran et al. 2009 as a starting point). How might some of the tools and technologies, such as those of crisis mapping and crowdsourcing, be used to capture this local spatial knowledge to inform disaster mitigation?

RESOURCES NOTES

The *Community-Based Vulnerability Assessment* at http://www.mdcinc.org/sites/default/files/resources/Community%20Based%20Vulnerability%20Assessment.pdf (2009) is an excellent practical guide to doing social and physical vulnerability assessments at the community level and with a strong emphasis on mapping.

United States Geological Survey (USGS) "Earthquake Hazards 101: The Basics" at http://earthquake.usgs.gov/hazards/about/basics.php.

The US Census Bureau Census Reference GIS datasets at http://www.census.gov/geo/maps-data/data/tiger-line.html include reference GIS datasets such as census tracts, blocks, or block groups can be downloaded directly from the US Census Bureau. These datasets can then be used as the basis for creating thematic maps by combining them with various indicator datasets and downloaded through the American FactFinder website.

"Joining Census Data to Shapefiles in ArcMap" is a step-by-step guide on how to use census data tables in ArcMap using table joining functions at http://spatial.scholarslab.org/stepbystep/joining-census-data-tables-to-shapefiles-in-arcmap/.

The QGIS 2.0 Workshop provides a general guide to joining tables in QGIS at http://maps.cga.harvard.edu/qgis/wkshop/join_csv.php.

The USGS Earthquake Hazards Program at http://earthquake.usgs.gov/ is an excellent source of global earthquake data.

Place name points for the United States (which include things like schools and hospitals), can be downloaded from the Geographic Names Information System at http://geonames.usgs.gov/domestic/download_data.htm.

To learn more about the Weighted Overlay Tool in ArcGIS, see http://resources.arcgis.com/en/help/main/10.2/index.html#//009z000000rq000000.

For equivalent tools in QGIS to perform the analysis of developing a spatial index of vulnerability, see http://pyqgis.org/repo/contributed and MCELite : 0.1.2.

To learn more about the development of maps for flood risk assessment in flood risk management, see the FEMA Flood Hazard Mapping website at http://www.fema.gov/national-flood-insurance-program-flood-hazard-mapping.

REFERENCES

Arnold, Garth, Brett Carlock, Mike Harris, Adam Romney, Mario Rosa, Josh Zollweg, Anthony Vodacek, and Brian Tomaszewski. 2012. "GIS modeling of social vulnerability in Burkina Faso." *ArcUser* 15 (1):20–23.

Birkmann, Joern. 2006. "Measuring vulnerability to promote disaster-resilient societies: Conceptual frameworks and definitions." In *Measuring Vulnerability to Natural Hazards: Towards Disaster Resilient Societies*, New York: United Nations University Press, 9–54.

Chaudhuri, Shubham, Jyotsna Jalan, and Asep Suryahadi. 2002. "Assessing household vulnerability to poverty from cross-sectional data: A methodology and estimates from Indonesia." *Department of Economics Discussion Paper Series* 102:52.

Cutter, Susan L., Bryan J. Boruff, and W. Lynn Shirley. 2003. "Social vulnerability to environmental hazards." *Social Science Quarterly* 84 (2):242–261.

Cutter, Susan L., Joseph A. Ahearn, Bernard Amadei, Patrick Crawford, Elizabeth A. Eide, Gerald E. Galloway, Michael F. Goodchild, Howard C. Kunreuther, Meredith Li-Vollmer, and Monica Schoch-Spana. 2013. "Disaster resilience: A national imperative. " *Environment: Science and Policy for Sustainable Development* 55 (2):25–29.

Department for International Development (DFID). 1999. "Framework, Section 2.1." In *Sustainable Livelihoods Guidance Sheets*. London: Department for International Development (DFID).

Douglas, John. 2007. "Physical vulnerability modelling in natural hazard risk assessment." *Natural Hazards & Earth System Sciences* 7 (2).

Ebert, Annemarie, Norman Kerle, and Alfred Stein. 2009. "Urban social vulnerability assessment with physical proxies and spatial metrics derived from air- and spaceborne imagery and GIS data." *Natural Hazards* 48 (2):275–294.

Environmental Systems Research Institute. 2007. *GIS Best Practices: GIS for Earthquakes*, Esri, http://www.esri.com/library/bestpractices/earthquakes.pdf (accessed March 31, 2014).

Environmental Systems Research Institute. 2014. *About the ArcGIS Spatial Analyst Extension Tutorial*, ArcGIS Resources, http://resources.arcgis.com/en/help/main/10.2/index.html#//00nt00000002000000 (accessed March 31, 2014).

Federal Emergency Management Agency (FEMA). 2010. "Social vulnerability approach to disasters," FEMA Emergency Management Institute, http://training.fema.gov/EMIweb/edu/sovul.asp.

Frankenberger, Timothy, Kristina Luther, James Becht, and M. Katherine McCaston. 2002. *Household Livelihood Security Assessments: A Toolkit for Practitioners*. Atlanta, GA: CARE USA, PHLS Unit.

Fung, Vincent. 2012. "Using GIS for disaster risk reduction," United Nations Office for Disaster Risk Reduction, http://www.unisdr.org/archive/26424 (accessed March 31, 2014).

Kamp, Ulrich, Benjamin J. Growley, Ghazanfar A. Khattak, and Lewis A. Owen. 2008. "GIS-based landslide susceptibility mapping for the 2005 Kashmir earthquake region." *Geomorphology* 101 (4):631–642.

Kappes, M.S., M. Papathoma-Köhle, and M. Keiler. 2012. "Assessing physical vulnerability for multi-hazards using an indicator-based methodology." *Applied Geography* 32 (2):577–590.

Kemp, Randall B. 2008. "Public participatory GIS in community-based disaster risk reduction." *tripleC: Communication, Capitalism & Critique. Open Access Journal for a Global Sustainable Information Society* 6 (2):88–104.

Maguire, David J., Michael Batty, and Michael F. Goodchild. 2005. *GIS, Spatial Analysis and Modelling*. Redlands, CA: ESRI Press.

Mileti, Dennis S. 1999. *Disaster by Design: A Reassessment of Natural Hazards in the United States*. Washington, DC: Joseph Henry Press.

Papathoma-Köhle, M., M. Kappes, M. Keiler, and T. Glade. 2011. "Physical vulnerability assessment for alpine hazards: State of the art and future needs." *Natural Hazards* 58 (2):645–680.

Phillips, Brenda D., Deborah Thomas, A. Fothergill, and L. Blinn-Pike. 2010. *Social Vulnerability to Disasters*. Boca Raton, FL: CRC Press.

Schneiderbauer, Stefan, and Daniele Ehrlich. 2004. *Risk, Hazard and People's Vulnerability to Natural Hazards: A Review of Definitions, Concepts and Data*. Brussels: European Commission, Joint Research Centre (EC-JRC).

Tran, Phong, Rajib Shaw, Guillaume Chantry, and John Norton. 2009. "GIS and local knowledge in disaster management: A case study of flood risk mapping in Viet Nam." *Disasters* 33 (1):152–169.

United Nations Office for Disaster Risk Reduction (UNISDR). 2007. "Terminology: 'Resilience,'" UNISDR, http://www.unisdr.org/we/inform/terminology (accessed February 7, 2014).

United Nations Office for Disaster Risk Reduction (UNISDR). 2007. "Terminology: 'Vulnerability,'" UNISDR, http://www.unisdr.org/we/inform/terminology (accessed February 7, 2014).

United Nations Office for Disaster Risk Reduction (UNISDR). n.d. "Our Mandate," UNISDR, http://www.unisdr.org/who-we-are/mandate (accessed June 8, 2014).

United States Department of Homeland Security. 2013. National Mitigation Framework, http://www.fema.gov/national-mitigation-framework.

9

Special Topics
The Future of GIS for Disaster Management, Developing a GIS for Disaster Management Career, and Keeping Up with Current Trends

CHAPTER OBJECTIVES

Upon chapter completion, readers should be able to

1. understand selected special topics for Geographic Information Systems (GIS) and disaster management;
2. develop an appreciation for the future of GIS for disaster management based on the perspectives of people from academia, the private sector, national disaster management agencies, the United Nations, and local and federal governments;
3. discern specific advice on building a GIS for disaster management career;
4. understand how to stay current in the GIS for disaster management field using a variety of mechanisms such as organizational membership, conferences, journals and magazines, training education, and volunteer opportunities.

INTRODUCTION

Four areas are discussed in this final chapter. The first is GIS for disaster management special topics, which is presented to give you a sense of the breadth, variety, and potential of GIS for disaster management research and development activities. The special topics are by no means an exhaustive list. You are encouraged to use these special topics as a starting point to "think outside the box" as to what is possible with GIS for disaster management.

The second chapter area is the future of GIS for disaster management, in which we present the conclusions of many of the interviews conducted for this book. In particular, you will get perspectives on where the field of GIS for disaster management is heading in the next 10 years or sooner from the wide range of people who were interviewed for this book. These and other ideas are then summarized into a GIS for disaster management research agenda.

If you are a student of GIS for disaster management and you are interested in pursuing a career in this field, the chapter's third area provides extensive advice on building a career and finding a GIS for disaster management job. This chapter section is also based on the interviews of working GIS professionals that were conducted for this book. All of them have worked with many people in the early stages of their careers and thus have sound, practical advice to offer you. If you are currently working in the GIS for disaster management industry, the final chapter area offers some advice on staying current in the field based on a wide variety of organizations, journals, conferences, and other outlets relevant to both the academic and practitioner side of GIS for disaster management. Volunteer opportunities are then provided if you have time available and you are interested in learning about the disaster management field from a hands-on perspective.

SPECIAL TOPICS

Visual Analytics

Visual analytics is the science of analytical reasoning supported by interactive, visual interfaces (Thomas and Cook, 2005). Analytical reasoning is the process by which human judgment is used to reach conclusions based on a range of evidence within a set of assumptions (Cox, 1999, 1996). Analytical reasoning can often be visually supported to take advantage of human capabilities associated with vision and cognition (Larkin and Simon, 1987; Zhang, 1997; Zhang and Norman, 1994). Examples of this idea range from simple tasks, such as using one's fingers to count, all the way to complex visualization tools to support scientific inquiry. Thus, the fundamental idea of visual analytics is using computational tools to transform, modify, and process a wide variety of evidence and then present evidence in interactive, visual interfaces to support a human analyst in developing hypotheses, testing hypotheses, and ultimately making a final decision. Figure 9.1 illustrates a disaster management example of visual analytics ideas.

In the top left of Figure 9.1, news reports related to food shortages in Africa have been acquired through an RSS (Really Simple Syndication) feed, and the first story has been selected for further analysis. When the analyst clicks on this story, shown on the bottom left are relevant people, places, and thematic dimensions of the story highlighted using a variety of colors so the analyst can quickly determine if the story contains any potentially important items that warrant further investigation. The geographic map shown on the right of Figure 9.1 highlights places found inside the news stories using computer algorithms to extract geographical references from the text. Geographical references are extracted from the text and then rendered on the map to visually determine if any geographical patterns can be found from the various news reports. For example, if multiple news reports mention the same locations, these locations would appear as a cluster. Using a visual analytics system like the one shown in Figure 9.1, an analyst can examine multiple forms of evidence, such as computationally transformed text derived from news reports

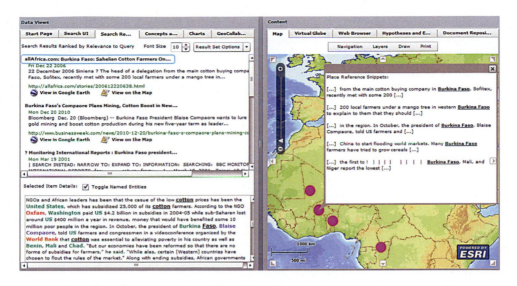

Figure 9.1 A visual analytics system. (Based on B. Tomaszewski and Alan MacEachren. 2012. "Geovisual analytics to support crisis management: Information foraging for geo-historical context." *Information Visualization* 11 (4):339–359.)

with interactive visual interfaces such as a digital map and a text analysis interface to highlight potentially relevant people, places, and organizations. Although not shown in Figure 9.1, additional visual interfaces supporting different forms of evidence can also be included such as event timelines, data charts, concept graphs, document repositories, social media feed, and images. Ultimately, combining multiple forms of evidence through visual interfaces can potentially help an analyst find interesting patterns, trends, or items of interest for further analysis that might not otherwise be apparent by examining the forms of evidence individually or in nonvisual formats.

Big Data and Disaster Management

A concept closely related to visual analytics is that of *big data*. Big data can be thought of as datasets where the volume (overall size), variety (different forms such as social media, imagery, geospatial, email, etc.), and velocity (speed at which it is produced and analyzed) challenges the thinking and existing techniques surrounding these issues (Jacobs, 2009; Gartner Inc., 2011). Representative examples of geographically oriented big data in disaster management include

- millions of location-tagged tweets and other social media artifacts such as people using their phones to take pictures of disaster situations generated in almost real time during major disasters;
- multiple decades' worth of daily satellite imagery collected through platforms such as Landsat that can be used for analyzing land use change for disaster mitigation and climate change adaptation research;

- massive amounts of imagery collected from heavily impacted disaster areas over a large geographic region that need quick analysis for decision making; and
- massive amounts of volunteer geographic information created through mediums such as OpenStreetMap.

The use of big data derived from social media and broader grassroots approaches in disaster management practice is still in its infancy. Although big data has been a subject of academic research for many years and disaster management organizations such as FEMA have begun exploring the use of big data, challenging issues still remain around trust and reliability of the data received, ensuring privacy of citizens, and developing better understanding between volunteer groups that create relevant big data and government organizations that use such data (Crowley, 2013).

In terms of technologies that can be utilized for processing big data in GIS for disaster management applications, many of the commercial and open-source technologies you were first introduced to in Chapter 3 contain advanced statistical and other spatial analytical methods that can be combined with geographic visualization for processing, analyzing, and making sense of big datasets.

You're also encouraged to explore technologies such as mongoDB (http://www.mongodb.org), which is a type of *noSQL* database. The difference between noSQL and relational databases (like those typically used by most commercial GIS databases discussed in Chapter 3) is explained in this quote from the mongoDB.com website (mongoDB, n.d.):

> NoSQL encompasses a wide variety of different database technologies and were developed in response to a rise in the volume of data stored about users, objects and products, the frequency in which this data is accessed, and performance and processing needs. Relational databases, on the other hand, were not designed to cope with the scale and agility challenges that face modern applications, nor were they built to take advantage of the cheap storage and processing power available today.

For example, mongoDB can be used to quickly and efficiently store massive numbers of tweets as they are generated during a disaster event and then quickly query, condense, and aggregate large tweet volumes for decision making using a map-reduce processing procedure (see mongoDB [2014] for technical details on how map-reduce processing works in mongoDB).

In general, big data will continue to be an important component of GIS for disaster management, and you are encouraged to conduct your own research and stay informed about broader developments with big data and how those developments can be related to GIS for disaster management.

Serious Games for GIS and Disaster Management

Serious games is the idea of games with a nonentertainment purpose (Michael and Chen, 2005). Geography-based virtual games have seen increased attention for teaching spatially oriented concepts such as resource management and human–environment relations (Ahlqvist et al., 2012; Cheng et al., 2010). Thus, the idea of serious games for disaster management is easily transferable to the disaster management domain given the fundamental spatial nature of disaster. Furthermore, as discussed in Chapter 5, scenarios and simulations can be an important aspect of disaster planning, and using gaming concepts such giving a score based on actions taken within a scenario can be an important metric to measure learning progress based on scenario parameters. As an example of these ideas, at the Rochester Institute of Technology we have been working to develop a serious game for teaching disaster management professionals about the capabilities of GIS and improve their overall spatial thinking abilities (Figure 9.2).

Figure 9.2 shows the serious game for disaster management spatial thinking. The scenario behind this game is that toxic waste barrels, as shown by triangles in the map, have washed up onto the shore of a major river. The game player then needs to make a series of decisions utilizing spatial thinking, as supported by GIS tools, as to how best to responded to this disaster. A key feature of this game is that it runs inside commercial-grade GIS Software (ArcMap 10.2), thus the game player can utilize real GIS functionality. However, the game player does not need to have technical skills to operate the GIS software, and can make GIS actions happen by simply pressing buttons, which in turn make the underlying GIS functionality operate, thus allowing the player to focus on making decisions based on good spatial thinking and not become distracted by GIS software operation. In Figure 9.2, the player has selected to buffer the barrels by 500 feet as an outcome of thinking spatially about the potential hazard zone of the barrels. The game then prompts the player to decide which data layer should next be buffered in relation to risk of population in terms of the barrel locations. Each action the player takes, based on three choices provided in each question (as shown in Figure 9.2), is assigned a score, with a higher score indicating the best action, and a lower score indicating a less desirable action. At the end of all the questions, the player is then given a final score that they can use to gauge how well they were

Figure 9.2 A serious game for disaster management spatial thinking. (ArcGIS software screen shot Copyright © 2014 Esri, ArcGIS, ArcMap. All rights reserved. Used with permission.)

thinking spatially and to measure improvement in spatial thinking with repeated scenario use. Ideally, approaches like the GIS serious game for disaster management shown in Figure 9.2 can be used to demonstrate the capabilities of GIS in an easy-to-use, yet realistic manner. For more details on this research, see Blochel et al. (2013); Tomaszewski et al. (2014).

Geographic Information Science and Disaster Management

Geographic information science is an interdisciplinary field concerned with the underlying theory and methods of GIS, questions related to the nature of geographic information itself and the impacts of geospatial technology on society and individuals, and the impacts of society and individuals on geospatial technology (Mark, 2003; Dibiase et al., 2006). Another way to think of geographic information science is that it seeks to develop the next generation of GIS technology through scientific inquiry in a wide variety of fields such as computer science, information technology, spatial cognition, statistics, cartography, and any other discipline relevant to GISystems. For example, the visual analytic system previously discussed is an example of a new technology developed from the perspective of geographic information science as the tool was examining nontraditional sources of geographic information such as text documents. Thus, if you become deeply interested in GISystems through operation of out-of-the-box software such as Google Earth or ArcGIS, I encourage you to take your interests to the next level. Help research and develop the next generation of new ideas, methodologies, and applications that can eventually be transitioned into a disaster management practice. Use tools such as Google Scholar (http://scholar.google.com/) to find out about the latest research in geographic information science, and develop technical skills such as computer programming for building new GIS disaster management technology based on your research.

THE FUTURE OF GIS FOR DISASTER MANAGEMENT

Interviews

Throughout this book you have met people from academia, the private sector, local government, international nongovernmental organizations (NGOs), the United Nations, and the US and German governments. As you saw, these people all had their own unique perspectives on GIS for disaster management. For most of the interviews, I asked the same question: "Where do you see GIS for disaster management heading in the next 10 years?" The following sections present their respective responses to this question. Pay very close attention to the items they mentioned as you will see several recurring themes emerge on the future of GIS for disaster management.

Jen Zimeke, PhD, Crisis Mappers (Chapter 1, Specialty: Crisis Mapping)
Where do you think the future of crisis mapping is heading?
In the beginning, conversations in the Crisis Mappers community primarily focused on data retrieval. How do we collect data from SOS text messages and geolocate and translate these data? How do we use data ethically and in a way that preserves the security and safety of individuals who have told us their stories? How can

we verify the content of the data we are collecting? Now, despite the continuing importance of these problems and questions, we are moving on to additional arenas of inquiry. The one that fascinates me the most is the area of data visualization and analysis. In other words, now that we have all this data, what can we learn from it? What are good questions that can be asked, and hypotheses that can be tested? And, coming from my own research and perspective, what can we learn from the data about *processes* themselves, like how disasters work, the nature of election cycles, or the dynamics of conflict? Can we develop a better way of understanding such complex dynamics, drawn from the micro-level event data we have collected and observed?

In short, I think that the analysis and visualization component of Crisis Mapping is going to become more and more important all the time, awash as we are in micro-level event data. I think we need innovation in the area of data analysis and visualization, and should strive for ever-more sophisticated treatment of these data. The study of complex dynamical systems, for example, may help us understand dynamic, endogenous processes, and how local, micro-level events can produce complex patterns and structures at the macro-level. In order to take the next step, we need to engage even more directly scholars, methodologists, and those working in diverse areas, ranging from those who do research in neural nets to complexity science to others who use machine learning or agent-based models for analysis, to GIS researchers and those working on the way in which complex games can be leveraged for disaster response.

I also think that a growing part of the community in the future will be concerned with understanding intuition. Just what, exactly, is intuition? How can we think about and better understand expertise and knowledge, especially the kind of knowledge that comes from deep experience? How can we capture the hidden data and analysis assumptions lurking behind a "hunch" or a "gut feeling"? It will become increasingly important, I believe, to interview, work with, and analyze decision-making processes by learning from policemen and firefighters with 30 years of experience, as well as 911 responders, disaster managers, and our war veterans, to name a few. When the fire captain, for example, sees how a given wildfire is moving and progressing over time, their hunch about whether or not the fire is going to jump the road, and thus whether or not to continue fighting the fire from this position or pull the team back, is a life or death decision, and must be made in a moment. To be wrong could cost many lives. How do they decide as they decide? Intuition is the impossible calculation, the mind attempting to maximize an equation with many variables, using the hundreds or thousands of prior incidents to help inform their decision. The mind doesn't necessarily know what it is grokking, or how, or even that it is doing so, and it doesn't have time to really absorb the data, consult area experts, and concoct a plan of action. The decision maker typically has key variables available to help inform this decision, including current wind direction, velocity, and speed, and real-time views of the fire landscape, but the decision about the complex dynamic inherent to the fire, and whether or not it is reaching a phase transition, must be made, and there is no time. What we need is a way to help bottle up some of that intuition and expertise, so as to understand how breathtakingly exquisite decisions are made on the fly, tamed at the intersection between art and science.

I would also love to see small teams of different people emerge to work together during a disaster and then disband afterward. I could imagine being physically in the same room around the map and around the dataset, together with eight or so different colleagues. One might be a statistician or someone that does agent-based modeling together with a linguist and an area specialist that understands the actual on-the-ground situation very well. The software or IT specialist would also be on hand. You could have all these people looking at the same data and then really helping make policy decisions and in a better way. So, what I think we need are these really small, micro interdisciplinary teams that can come together and that are handpicked according to different niche areas. I don't know the best way to create such a group, or who should do the creation, but it would be an important next step. We all can't be experts in everything, after all.

Thankfully, there are always voices in the community that remind us about the importance of affected populations and reach-back. After we've visualized and analyzed the data, what next? How do we directly involve affected populations in the humanitarian response in the ground, and in the data analysis itself? And how does all of this information help disaster managers with their important work? Are they getting the data that they need or are they just getting even more overwhelmed by too many data, and too much analysis? I think another domain of crucial importance to this area of inquiry revolves around how to ensure we don't overwhelm responders who are already busy enough during a disaster. How do we get that one gold piece of information you may care about into your hands, and not bother you with the rest?

It has also been very exciting to watch how the very definition of what *crisis mapping* is has emerged as a collaborative conversation and continues to change over time. We always wonder: what are the limits of its domain? To what can this concept be applied? For example, when Patrick Meier and I started Crisis Mappers Net, we had an idea in our mind about what we thought crisis mapping was, but then when we saw our friends from MIT Media Lab and Grassrootsmappers mapping the oil spill in the Gulf of Mexico using handmade kites and balloons for aerial photography. I remember thinking, "that never would have occurred to me. This is genius." So, we must remember there are as many different ideas about what crisis mapping is, and also how it could be applied, as there are people. I am excited that a new crop of young leaders have emerged very quickly in this field, and they are standing up and participating and pushing the field in new directions that we never anticipated, and are doing it better, in my view, than some of the old hands ever did in the first place.

The crowd is starting to learn that they can help act as an information filter, to help make sense of all of this big data. So far, the crowd has helped process and clean incoming SOS messages, tagged photos to help narrow the search for a missing plane, conducted damage assessment on homes after a hurricane, and used satellite imagery to get appropriate estimates for the number of refugees in a camp below by counting blue tarps, to name a few. However, there is still untapped potential to tap a wider global crowd that wants to engage, but doesn't necessarily know how. I think we are missing many opportunities for really leveraging the crowd, and enabling millions, instead of the hundreds or thousands, to participate. I look forward to the not-too-distant future in which this problem is greatly ameliorated.

In terms of the future, I think we are speedily moving toward two very different kinds of worlds. In one scenario, technology, in general, continues to be an ever-present part of our lives. Our immediate environment will change, as user interfaces change, social practices around the use of technology change, and the internet-of-things takes hold. User interfaces will probably allow us to view data as a visual layer over the real, actual environment, so, for example, I will walk down the street and see hovering over your head "Brian Tomaszewski," in case I forgot your name, and additional information as well, such as "the last time you talked to him was 2012 in Rochester," or whatever I need to help me with the upcoming interaction. So, in the future, we will have a vastly different way of interacting with data, as the distinction virtual and real, data and physical reality, begins to blur and then vanishes. This new reality opens up interesting questions, then, about what maps are. Will the spatial environment still be the primary way in which we organize all of the data? What will my niece and nephew's generation think a map is? If data is at your fingertips and GPS directions emerge from a thought or a tap of the head, what does that do to our conceptualization of what maps are? Along these lines, you have to wonder whether the term *crisis mapping* is simply a term that will vanish over time, or whether the term itself will persist and take on new meaning, but in a different direction.

Of course, the world envisioned above may never arrive. I could equally imagine an alternate world in which we overload the carrying capacity of not only the environment but also the Internet and everything that the Internet depends upon, including the power supply, and the belief that the Internet is mostly reliable and secure. If warfare increasingly moves to a cyber domain, and if the Internet is increasingly subject to attacks from hackers as well, then the future is one in which we should expect intermittency and unreliability to be the only constant. In this scenario, I envision more and more power outages and unreliable, intermittent, and insecure Internet connections and power service. Blackouts and brownouts will be commonplace and disruptive. The likelihood that the whole Internet is going to turn off forever or that power is going to go away and that we're going to have a collapse of humanity actually seems quite low, and even if it happens, it seems there is very little we can do about wholesale collapse in any case. However, I do think we should prepare ourselves for local collapses and unstable, unreliable online worlds marked by intermittent service, dubious connections, and heightened uncertainty.

Of course, we don't really know what the future holds. But back to your question: What will crisis mapping be like in the future? Crisis mapping could become utterly obsolete. The future could render it irrelevant. It could also be absorbed and completely integrated into modern life. Who knows?

Anthony Robinson, PhD, Penn State (Chapter 2, Specialty: Cartography)

Where do you think the future of disaster mapping is heading?

I expect that some of these design problems will be solved and that we will be looking at more intelligent systems that make it possible for nonexperts to make appropriately designed, effective maps—not just maps—and in a lot faster manner than we currently can do this. I also think the future of disaster mapping is going to have to reorient itself around a fundamental principle that some portion, if not the majority of the data that you're going to represent, will be coming from unofficial sources, in real time, and in extremely large volumes. We haven't yet developed methods for dealing with any of those factors in a maximally effective way. We are already trying to wrap our heads around Twitter and other social media, but try to imagine what it will be like twenty years from now. The notion we would be waiting for the US Geological Survey [USGS] to provide a model for the impact of population from an earthquake will be quaint at that point. We'll probably be remembering that "Wow!, remember when we had to wait for that?" because now we have a model that tells us, based on all sorts of other media that people are just generating by virtue of their normal interactions, and we will know that very specific things have happened from that interaction data. I think we will be smarter about recognizing anomalies and signatures of interaction—even from partial responses in places that have incomplete coverage. For example, you've got a certain number of people in an area who have a certain kind of device or use a particular service. I think the pervasiveness of that kind of stuff is probably going to increase substantially, and it will reach a point where we can actually detect silence as having serious meaning. That will be a tipping point. We will say "the chatter has died down in a place" in such a way that we can know that something serious has happened there. I think that will be a point at which we can reorient a lot of our immediate disaster response mapping around ingesting these real-time sources of contributed media. Because that's where we're going to pick up that (lack of) signal. Certainly other new data sources will be formalized and authenticated, as I'd expect that governments will continue to create sensor networks to also augment contributed media with somewhat less biased information. I think cartography will have to respond through the development of representation methods to reveal the dynamic signatures of large and ever-changing datasets of all kinds. Right now, we can make good static snapshots of these things and given the appropriate amount of time, you can also design them appropriately to widespread consumption. So if we can overcome the time challenge, we also have to overcome the dynamic challenge which is, how do we show disasters as processes unfolding and how do we show that through mapping? We need to do that instead of just starting a map at time zero, slamming new data on top of the old data. We don't have good ways yet to rapidly build meaningful map animations or to show multiple ways of comparing time with geography. There are plenty of good examples of dynamic representation techniques in the GIScience research community, but they haven't made their way yet into usable and useful disaster mapping systems.

Alan Leidner, Booz Allen Hamilton (Chapter 4, Specialty: Private-Sector GIS)
Where do you see GIS for disaster management in the next ten years?
My thinking is largely rooted in what worked and what didn't work during the responses to 9/11 and Sandy, two disaster events where I played coordinating and leadership roles. I learned that even if you created a great data repository prior to a disaster it doesn't mean you are in position to effectively provide the response community with the information it requires. Other things need to be in place. How much threat and vulnerability analysis was done beforehand and was that information used as the basis for taking preventive measures? Was the GIS data, including the use of weather predictions and storm modeling, used as an integral part of exercises so that the response community could become familiar with their use? Was there effective communications with first responders in the field and with the populations affected by the disaster and did the means exist to turn the information from the field into actionable intelligence?

New York City, arguably, has the best municipal GIS data in the world. Over the years, tens of millions of dollars have gone into building the city's enterprise GIS system. I also think New York City's GIS community did a great job responding to both 9/11 and Sandy. I'm not sure any other municipal GIS could do better. But I continue to think we might have been able to do a better job with improved preparations.

When a disaster strikes, the disaster itself generates information that can dwarf in scale anything you've collected in the past. Also, the data generated by the disaster is different. It's data about the changes caused by the disaster and that damage will be continuously evolving over time so it is critical to keep up with. Following 9/11, imagery and other remotely sensed collection efforts were flown daily for weeks on end. We rapidly ran out of space to store the data and did not have the capacity to analyze all that we wished. Also, there are many different streams of data coming from every corner of a disaster area. This data about critical supply chains, infrastructure systems, and the information collected by the field staff of utilities, public safety agencies, and service agencies is essential to managing rescue and recovery efforts. Having the preevent data is essential too for change detection, but by itself it doesn't help you understand and react to the real effects of the disaster. Following a disaster you start fresh every day.

We all know that information is critical to every human activity. Without information, without knowing what's happening where, it is impossible to act with any effectiveness and with any semblance of coordination. Ever since 9/11 I've understood that information organized by location was key to getting information needed both for preparing for and responding to a disaster. A disaster is an enormously complex event and dealing with it requires so much data from so many different sources over an extended period of time, that only GIS, with its information integration and visualization capabilities, can possibly manage the task. But this means that in order to optimally utilize GIS capabilities, the entire response community from decision makers to first responders must organize themselves to manage information better.

My impression is that for most Emergency Operations Centers the use of GIS following a disaster event is largely an ad hoc process. Lines of communications and data sharing are hastily put in place after the disaster strikes or a day or so before a predicted event occurs. Ties need to be built with major operational agencies on the fly. Communications with the field and with the public continues to be primitive and inefficient. GIS staff are put into a position where they must design and build information sharing mechanisms right in the middle of an event, because the game plan for acquiring all of this data and doing something useful with it in the real time does not exist even to this day. Everyone *talks about it.* No one really does it, although there are some promising federal tools being built such as the Virtual USA and the FEMA GeoPortal, which need more attention paid to them.

We learned a lot from 9/11 and as a result the response to Sandy was better, but it left much to be desired. And, in truth, even though the GIS managers in the region created a task force two days prior to landfall, we still fell rapidly behind events on the ground. For example, we were blindsided by the fuel shortage and initially had great difficulty understanding the fuel supply chain in the region. We also did not fully understand how badly electric power supply could be effected by the Sandy storm surge, nor had we done the analysis needed to understand the cascading effects that might occur when facilities located in flood zones tried to switch to backup power generators, which were placed in locations that were also vulnerable to flood waters.

For me, this was best exemplified by what happened to the electric substation and power plant at East 13th Street on the Lower East Side, near the East River. The East 13th Street facility channels power into Manhattan through feeds from large Queens power plants, and from power plants in northern New Jersey. The East 13th Street electric power complex is enormous: two entire huge super blocks totaling about eight acres. And it's within a flood zone. So, if you looked at the geospatial data and did a little back-of-the-envelope analysis, you could understand just how strategic this facility is and see that this facility in particular was going to be at risk in a major storm surge event. As Sandy approached the New Jersey coast, NOAA [National Oceanic and Atmospheric Administration] predictions indicated that there was a chance that the storm surge would set a record along the NYC Metro Area waterfront. But by then it was too late to sufficiently harden the substation. Explosions within the facility could be seen from across the East River in Brooklyn, and power to Manhattan south of 34th Street to the Battery was knocked out for four days. This included the entire downtown financial center, a portion of the midtown business district, and countless residential, institutional, and commercial buildings.

So here you have a situation where geospatial data and analysis indicated that of all the city's electric infrastructure, the East 13th Street substation could be in greatest jeopardy and its failure could have the worst effects, and yet there seemed to be no one who paid attention to this information so that action could be taken well in advance. As a result, billions of dollars in losses were sustained. And I believe that this is a common problem. You've got smart GIS personnel who point out major risk, but the information is not taken seriously.

Similarly, a lot of the underground transportation facilities in lower Manhattan, such as transit tunnels and station, and the entrance to the Brooklyn Battery tunnel were flooded at a cost of billions. They were also located in known flood zones. And the GIS analysts who had looked at this situation knew that if there was a substantial surge, those facilities would be flooded. However, this knowledge never got translated into protective action by the responsible agencies and organizations. The intelligence offered by GIS is significant and accurate, but GIS experts are rarely offered a place at the table when decisions are made or funding allocated. It was that gap between GIS knowledge and taking effective action in advance that really clobbered us during Sandy.

What you just described at great length is a common problem that's existed for a long time—dissemination of getting actionable information into the hands of the right people in a way that they can consume it easily. How do you take a complex analysis and boil it down to a one-page or a one PowerPoint slide or something? From the perspective of the private sector then, are the private sector consultants the ones that are producing good analytical products or is there a disconnect between the contractors getting good analytical products to the government officials?

I believe that a lot of the good analysis is being done by government GIS personnel and their contractors who are often just not being listened to. At the same time, I'm certain that there are good engineers and mapping specialists working for Con Edison and other utilities and private companies. I'm sure many Con Ed technicians understood that their infrastructure was at risk, because there were frantic efforts to shore up the 13th Street substation and other vulnerable facilities just prior to Sandy making landfall. More generally, in these kinds of cases, if someone with a technical background and GIS analytic capabilities says a facility is really vulnerable and we should do something about that in advance, the business decision to spend the necessary funds is often the roadblock. And many managers are just not convinced that they ought to do that even though all the predictions and all the models say it needs to be done and it will be cost effective in the long run. So, the power of GIS to influence these kinds of financial decisions is still in its infancy. And it's something that, I think, we're all working to improve.

What do you think it would take to get there? What would you—is it possible, ever be possible to really kind of break the disconnect and really get a good flow of information to the right people at the right time?

I think that what you're doing, actually, is an important step in the right direction. Your willingness to discuss these issues in the classroom and at professional events, and then to write about them is absolutely critical. The academic community is very important because they have the freedom to speak the truth without the fear of repercussions. And I think that one of the problems of working either in government or industry is that you've really got to be careful about what you say and you can't push an issue too far without putting your job at risk. So, there's a muffling effect that academia can be free of. And I think we look to higher education, to the professors and the students to actually raise a bit of a hullabaloo about this stuff. I also think that efforts of professional groups like the NYSGIS Association and regional groups such as NYC GISMO can be very helpful. Both NYSGISA and GISMO hold frequent webinars at which there is a free exchange of information and opinion. This may not always solve the problem but it gets useful information out in the open. So, maybe this is a ten- or twenty-year project. I don't know how long it will take, but I think it's doable. But it's just going to mean a lot of good people are going to have to push for a fairly significant amount of time to make change. I would also add that I remain optimistic about our political leadership. We've had some very strong support from city leadership in the past. I know some of the officials who recently took office in New York City have backed the development of GIS for many years. I think they will be receptive to what the GIS community has to say.

I'd also like to say something about special-needs registries. There are jurisdictions across the country that now compile listings of vulnerable individuals and map them within the service zones of their local fire or police stations. Also identified are the specific kinds of events that might threaten these individuals. So, for example, people whose well-being depends on electrically powered equipment would be on the registry so that if there's a blackout, first responders would know who might need assistance. If people live within a flood zone or along coastal areas and they have limited mobility or resided in a particularly vulnerable house, these registries would be used to inform evacuation efforts. And it's all driven by GIS. You have to know where these people are located: *the GIS component* of it. Well, what happened during Sandy was that quite a number of people drowned in New York City. They were mostly elderly living near the waterfront, within mapped flood zones, in basements or first floor apartments. Once the surge hit there was no escape. A number were found floating in their basements. So, here's an instance where a registry enabled by GIS location data, if it had been set up in advance, might have saved some lives. I now understand that NYC has recently authorized the creation of such a special-needs registry system. [See http://www.thenewyorkworld.com/2013/06/13/disaster-registry.]

I think it is also important to note that GIS is being used by New York City and many other jurisdictions on a number of lifesaving applications. This includes the 911 emergency response system that dispatches police, fire, and EMS personnel and the Compstat system that analyzes crime patterns and supports the design and implementation of crime reduction strategies. New York City has invested millions of dollars in a comprehensive and highly accurate street map and address database, that is at the core of these applications, which has contributed to the saving of thousands of lives. Yet most people do not understand the vital role played by GIS in making these applications effective. We need to make sure that the public understands the lifesaving role played by GIS and creates a demand for its greater utilization. So, in part, it's all about public education. And your book is another element of the strategy to bring GIS to peoples' attention.

One other thing I would like to add. I've started to use the expression: "from crowd to cloud" to represent the new data collection and synthesizing capabilities that are now coming to the fore. For example: in a disaster you want to collect information from the field, across a large area, from a wide variety of first responders, and from different agencies and organizations. This information needs to be combined with information from citizens and with data collected by sensors located in the disaster area or captured by airplanes and satellites. Then all this data needs to be analyzed and turned into useful intelligence products that are customized to meet the varied needs of the response community and those affected by the disaster. This is a huge technical challenge, but it is now becoming doable. Unfortunately, I really don't see government moving as quickly as it might towards this kind of a comprehensive, integrated, real-time data approach. True, there are small steps being taken in this direction, but there is an absence of an overall strategy. There is a huge opportunity to harness these capabilities to save lives. And it's going to happen. I'd just like to see it happen faster.

Antje Hecheltjen,* UN-SPIDER (Chapter 4, Specialty: Remote Sensing)
Where do you see GIS for disaster management in the next 10 years?
There's a lot of potential for much more near-real-time data access—not only of satellite but also of in-situ data. Specifically, I am thinking of the sensor web technology. I think this is the future. The data policies are changing, also in Europe. In the US there is a long tradition of very open data policy such as the opening of the Landsat archives,[†] but in Europe it was much more restricted. It's opening up now with the new Sentinel missions.[‡] Another topic that will become more important in the future is near-real-time processing of data to derive products for monitoring and early warning but also for response. Now such data products are already available—partly in experimental mode—with low spatial resolution, for example, for flood or drought monitoring. With still increasing computational power, new sensors in space and access to in-situ measurements, such services may become available operationally in future with higher spatial resolution to support disaster and risk management.

Michael Judex, PhD, German Federal Office of Civil Protection and Disaster Assistance (Chapter 4, Specialty: Federal Government GIS (Germany))
Where do you see GIS for disaster management in the next 10 years?
I think GIS will be much more service oriented. What we have seen is the shift from file-based storage to geodatabases. Then, we move forward to services because they're much more convenient to transport information during a crisis. I think this trend will dramatically increase because having standards to transfer data and information to other people and institutions is so useful. And as the services can be used via every telecommunication channel that provides Internet access, they can also be used by a satellite connection, for example. Interconnected services will be, I think, one trend. And then, I think geo information as such will be much more integrated in other services and also in devices such as tablets

---

* The views expressed herein are those of the author(s) and do not necessarily reflect the views of the United Nations.
† Landsat website, http://landsat.usgs.gov/.
‡ European Space Agency (ESA), "Copernicus: Observing the Earth," http://www.esa.int/Our_Activities/Observing_the_Earth/Copernicus/Overview4.

and other mobile devices. We're seeing much more firefighting professionals and technical relief agency staff members using such mobile devices. So, I think that will be an important part. One other point that will be important is that geoinformation will be becoming more and more a mean for the communication to the public. In Germany, most communication to the public is text based at the moment, but I think especially for risk communications (but also for crisis communication), maps are much better suited and able to transport complex risk information and, ultimately, also complex crisis information. So, that's a logical consequence from there. Additionally, I think a real challenge will be the meaningful use of the exploding amount of sensor information. We have more and more network-accessible sensors, such as automatic sensors like river gauges or traffic sensors but also human sensors that provide information. Think about the volunteered geographic information community. Talking about sensors, unmanned aerial vehicles will be one major trend I assume. That will be a real challenge to integrate all that information and to transform raw data into information and put them altogether on a platform that is able to support the decision making. The challenge will be to filter and aggregate the information flow to the really meaningful piece of information that is relevant for crisis management decision making. Finally, I think the use of models and simulations to look ahead, to look into the future, will be one future trend.

**Scott McCarty, Monroe County GIS (Chapter 4, Specialty:
County Government GIS (United States))**
Where do you see GIS for disaster management in the next 10 years?
I think in the next 10 years I see GIS becoming a major tool for anybody in emergency management and public safety, more so, on the ground level. I think right now we try to push out this GIS technology in the office or in the emergency operations center, but I really think that in the next 10 years you're going to see it out in the field more. We kind of do that with our mobile technology vehicle [discussed in Chapter 1]. But with smartphones and iPads, we're looking to build applications that can run on them so that any police officer or fireman or EMS person can look at their phone or their iPad and pull up the important information to help them while they're actually in the field. I've been doing the GIS thing for about 16 years. It's really probably only been in the last maybe 8 or 10 years tops that it played a major role in disaster management. When I first started with it, there was an EOC, but I just don't think there was any GIS in it as far as I can remember, or at least we weren't involved at that point. It really wasn't until we consolidated services in 2000 and maybe a couple years after that we started to kind of get our foot in the door as far as being involved in all these exercises and preplanning. I think from now going forward with what we have with the mobile unit and what we're going to build for our web-based applications and mobile device applications, I think it's just going to continue to grow. The technology continues to change every day and we keep our eye on that. There's certainly things we can do now that we couldn't do 10 years ago. When I started, command line GIS was on its way out at that point in the late '90s. We were using ArcView 3.2, which was just, it seems like, light

years behind where we are now. We're really keeping our eye on what Esri has to offer and also what other software companies have to offer. We look at what other states are doing, or other counties, or even towns to see what they have going on. If they're doing something that looks like it could be a benefit, then we'll see how we can apply that to what we're doing here in Monroe County.

Lóránt Czárán,* United Nations Cartographic Section and Office for Outer Space Affairs (Chapter 5, Specialty: Remote Sensing International GIS Organization, United Nations)
Where do you see GIS for disaster management in the next 10 years?
One important aspect will be data and services, and the efforts of various organizations in providing more access to their data holdings through web services, web map services, so that access to data is faster and easier. Related to this point, certain things where I expect a little bit more progress than in the past ten years is access to satellite data and good mechanisms for that access. We are making progress, but more can be done. We kept trying and then looking into funding mechanisms that are on standby for specifically licensing postdisaster satellite imagery for wider use, in the context of UN-SPIDER, of course. Those things didn't work for various reasons. Maybe donors and political levels were not fully convinced of what we were arguing for or maybe we could have done a better job in explaining what we—or why we need this. But, nevertheless, we are not there yet. So, those are a bit of a disappointment.

In terms of what I see happening next, I still hope and think that in the next 10 years we will learn from the mistakes of the past, after having captured all these repeated lessons learned from situations like we had in Myanmar or in Haiti or even with the Indian Ocean tsunami of 2004. In all these major disasters, the same problems seem to resurface again, with little improvement. So, I'm hoping that we will finally learn from all this and we would come closer in terms of collaboration. We would have mechanisms where there would be enough resources for a vast acquisition of critical data, especially satellite imagery or aerial imagery when something happens, data that would be licensed for wide consumption by all involved experts responding. I'm looking also at much wider use of aerial drones where situations allow for the more flexible collection of similar imagery data and maybe more flexibility in terms of what types of sensors could those drones carry. So, not your typical video camera or optical photo camera of today, but perhaps SAR [Synthetic Aperture Radar] instruments and other sensors that could be easily flown on drones as well or lighter to develop much higher resolution, much more accurate and much more penetrating data. I know when it comes to radar, hopefully cloud cover and forest canopy will not be an issue anymore when radar is more widely used on drones and not only from satellites or planes, for example. So, I'm hoping that we are going in that direction, that there's more flexibility in terms of collecting data and making that data available very fast after any disaster. For example, making sure that if data is purchased, it is purchased with a license allowing more wide sharing, or that

---

* The views expressed herein are those of the author(s) and do not necessarily reflect the views of the United Nations.

it's simply provided at no cost. We have mechanisms, such as the EU Copernicus program, that are stepping into the picture with provision of free satellite data, even if it's not very high resolution, but we are also talking radar besides optical, all of which will come gradually in the next 5, 6, 7 years. This would, of course, revolutionize the way the access to data is being seen or granted. We would still need very high resolution for many situations, and we still need to work closely with commercial partners, but I'm hoping that learning from what we had so far we will have the existing—we will have the means and the resources to also set up the partnerships with those commercial providers to fill in the gaps and to provide us quicker access to their data. Then, as well, since everything is moving to the cloud now even in terms of geospatial and we see the trend, having all this data delivered through services and so through the cloud rather than what we did even 10 years ago or with Haiti in 2010 (such as shipping hard drives, waiting for 3 days until they get to the field office, or downloading data during the tsunami in 2004, spending 2 weeks to download 700 gigabytes through the Internet and then running it on one or two computers to pansharpen and cut into 10 square-kilometer files, etc.). Those things should be of the past.

Hopefully with all these services, cloud computing and storage, the increase of bandwidth globally, the availability of bandwidth in many developing countries as well, hopefully data will make its way much faster to the end user on the ground and in better quality and with much more flexibility in use. That's what I would love to see and I'm actually hoping that we will get there in the next 10 years.

Then, in the same time probably some higher-resolution elevation data and other base data that could be of use to everybody would be developed at least or funded from various resources and made fully available to those organizations who need it in their work. Additionally, seeing private sector organizations or companies such as Google or Microsoft or Esri investing more in disaster management–related preparedness or having their own crisis management or disaster management teams also means that they channel a part of their investments into improving their efforts in this domain. To me, that is also something that will lead to a lot of advantages and improvements in the next 10 years because their services, their web services, their data holdings will all be geared towards better support in disaster management too especially as they have their own internal initiatives in that context. So, their technologists will also allow us for all that flexibility and access, faster access, better access that we need and that we talked about for so much time. So, that's where I see things, hopefully, and from my perspective, I think we are a good bunch of experts in all these organizations. If we would have better data access, faster access to postdisaster imagery, these things, I think, would make a big difference in making the point about the use of geospatial data when it comes to disaster management.

I think a lot of people are still not believing in the use of GIS for disaster management or that they have low expectations, especially because of the time it takes, but I'm convinced that with these developments that I mentioned, time will be less and less an issue. Some of the services and availability of data will be there

6, 12, 24 hours after something happens, much faster than anybody can be on the ground, and at a high spatial resolution that we are hoping for. For example, so we can easily and very, very accurately pinpoint areas to evacuate before a flood would come or even before a tsunami hits a coast. So, that's where we are. That all requires high-resolution data, high-resolution accurate elevation data, things that we don't really have today, but hopefully in 10 years will be different.

In my opinion, the problem now is that I put a nice satellite map on my wall and that's it and I cannot tell you in 6 hours after the rain falls upstream somewhere that certain areas will be flooded downstream or in that city, which is basically trying to predict the future. But those are exactly the kind of data and services we need. And for that to happen, we have to get out of habits such as reusing 20-year-old datasets at small scales. We need to move on. If that costs money, we need to make that investment. The same way as tents costs and biscuits costs, data costs too. Don't expect data for free because it comes too late and too slow. You pay for it. It comes immediately and it's useful. Same way as you have to pay for those tents or biscuits because no vendor will give it away for free. It all costs and somebody pays for it. We need to accept that data has a cost too and that an investment in data means everything works better, even the distribution of the tents and the food, simply put.

Today, decision makers are not in a position to strongly support or defend such investments, still. Every time you talk to one of the more senior ones, they will say, "I don't really see the benefit or the advantage of GIS— yeah, it's useful, great maps, but I can't invest in it or I can't take it seriously." Once this mindset changes, things will be different too. I'm hoping that with the increase of both bandwidth, cloud services, accuracy, scale resolution of data, we are getting closely—or slowly there. But a lot of it will depend also on the people in the different organizations and their willingness to collaborate because if everybody is just holding things for themselves and just funding in different directions and trying to capture all "attention and glory" and raise more funds than others, if this is the spirit that will continue, then disaster management itself will not be anything more than any business.

David Alexander, US Federal Government (Chapter 7, Specialty: Federal Government GIS (United States))
Where do you see GIS for disaster management in the next 10 years?
I think from the trade-craft side, I think it's becoming more and more specialized. From the technology side, it's becoming more and more consumerized and part of our DNA. It's like the weather. We pretty much don't do anything until we check the weather. I think you're not going to do anything without understanding the context of location or thinking about where and what's going on with where you're going.

I think the other trend that you're going to see is a rush towards democratization of GIS and situational awareness. Think of it from a sensor perspective—which is a good thing. We can't throw manpower at homeland security. We don't have enough forces to really secure the nation without recognizing the contributions our citizens can have. That doesn't mean we want our citizens to spy on anybody.

But under different scenarios, they now become a valuable sensor that can report information back to us. Whether they're leveraging something like, Did You Feel It*?, and they're reporting that they felt the tremor—and using that crowdsourced information as an input into the ground centers and calibrating our understanding of that event. Or, for example, in suspicious activity reporting and using that to identify if there any patterns that are occurring around vicinity and context.

Then, starting down the path of more citizens reporting more rapidly on things like did they suffer damage or experience a flooding, in most cases, citizens aren't going to mislead during times of duress intentionally. They're not looking to falsify information. They're looking to contribute and assist their neighbors and their community. I think you're going to see a rapid pace towards democratization, that means crowdsourced, and recognizing that sensors are human elements and not just technology elements. They're not just cameras out on the ground. They're not just biometric sensors or other sensors that are reading temperature and velocity and maybe chemical affluence, but they're also our people, the people component of our globe and the nation.

Research Agenda

The following is a suggested research agenda for GIS for disaster management based on perspectives derived from the interviewees of this book and other research conducted while creating this book. It is intended to be both an agenda for academically-oriented research, but also to give ideas for research and development of new, practical applications of GIS to disaster management.

1. Develop new methods to gather, process, transform, analyze, integrate, and curate greater volumes of "traditional" geographic information such as satellite/imagery, raster, and vector datasets along with new forms of geographic information increasingly important to disaster management such as big data, open-source information, social media, and crowdsourced information. Although social media and crowdsourced information will likely continue, not everyone will be using social media and providing crowdsourced information during a disaster (let alone after a disaster has passed). These data streams will not completely replace traditional geographic information. A balance between using both information types should be developed to take advantage of as many forms of information available from as many different sources as possible, while keeping information use grounded in relevant disaster management practice and culture. Best practice case studies based on combining traditional and nontraditional geographic information and technologies for disaster management should be documented and disseminated to academic and practitioner communities to help advance the state of the art.
2. Develop new forms of data dissemination services and examine how existing data dissemination services can allow larger and diverse amounts of geographic information to be more easily and quickly shared with a wider range of disaster

---

* USGS, http://earthquake.usgs.gov/earthquakes/dyfi/.

management actors on a wider range of technology platforms. Going forward, traditional desktop computing environments will be further replaced by tablet computers, smartphones, and smaller screen devices connected through the Internet. Development of new data dissemination services must also account for low- to no-bandwidth situations such as those in the developing world or in disaster situations of intense magnitude where critical communication infrastructures are destroyed or nonexistent until recovery processes of unknown time length can replace them.

3. Examine, identify, and document specific, relevant disaster management products that can be created by GIS and determine how those products can be used at the right time, by the right people, for the right situation, and on the right technology platform or display format. For example, the general public will have different needs for disaster products than working disaster management professionals. Paper-based maps, although effective in some cases, are time-consuming to create and require specialized skill so as to not mislead. Ideally, research and development in this area will draw upon user-centered design practice and usability engineering techniques to rigorously evaluate GIS product usefulness and relevancy (see US Department of Health & Human Services, 2014; Fuhrmann and Pike, 2004; and Robinson et al., 2005 for starting points to learn about user-centered design practice and usability engineering techniques).

4. Conduct organizational studies that identify the specific flow of geographic information across multiple levels of actors, including citizens, local responders, and government, where information flow bottlenecks occur and how this information can be curated, archived, and transferred to inform subsequent disaster phases. For example, information captured during a response can inform recovery, and then mitigation and planning.

5. Identify specific cases of the financial, organizational, and collaborative value of GIS for disaster management and present those cases to relevant decision makers who can enact policy and organizational change for further incorporating GIS into disaster management practice across all scales (see Joint Board of Geospatial Information Societies [JB GIS], 2013 as an example). Ideally, in 10 years, GIS will become so ingrained into the activities of citizens, disaster management practitioners, and decision makers that it will become an "invisible" technology much like the Internet or computers in general. In a sense, a trend like this is already starting in that many people are familiar with GIS as it is manifested in forms such as using Google Maps for directions, having a GPS navigation device in a vehicle, or using functionality such as the Facebook check-in feature. Ideally, these existing, familiar mechanisms can be tapped into for disaster management purposes such as people using offline, cached Google Maps to keep track of a wide range of *hyperlocal* activities in their neighborhoods once a disaster occurs. For example, knowing what business are open, where basic necessities like food and water can be obtained, and supporting community members.

6. Develop new educational and pedagogical approaches for teaching and raising the awareness of GIS capabilities and spatial thinking for disaster management to diverse groups of learners with diverse educational needs. Many disaster

management professionals only use GIS for a small portion of their actual work and do not require extensive educational training on how to use GIS. Furthermore, they may have the need for GIS in their work, but they may be unaware of what GIS can offer them, or they do not have the time needed to become proficient in operating GIS software like those discussed in Chapter 3. Free GIS data browsers such as Google Earth are a good start in this direction given their ease of use, but they are limited in their analytical capabilities. Ideas like the serious GIS game for spatial thinking or other easy-to-use learning environments that allow people to see the real capabilities of GIS using realistic disaster management scenarios should be developed to allow people to learn about GIS without interference from the software itself.

DEVELOPING A GIS FOR DISASTER MANAGEMENT CAREER

Interviews

In addition to providing insights about the future of GIS for disaster management, many of the people interviewed for this book have been working in the GIS for disaster management or related fields for many years and have seen numerous early-stage people enter the field. In the following sections, they offer advice on developing a career and finding a job in the GIS for disaster management field in relation to the area in which they work, such as the private sector, government, international NGO, or the United Nations.

Alan Leidner (Chapter 4)
What advice can you provide to someone new in their career that is working in disaster management for GIS and in the private sector?
Looking at the private sector, I believe there should be an increasing number of positions for GIS-trained personnel in corporate security and emergency management divisions. Many private sector companies including large financial services firms, transport companies, and gas, telecom, and electric utilities, have risk management divisions that often include an emergency operations center. In my experience, quite a number of these organizations don't utilize GIS personnel and do not understand how GIS information can help them. And the result is that they are unable to take advantage of all of the data and analytic resources available to them, especially from federal, state, and local governments. I think that as a GIS community we need to find ways to communicate with private sector firms and get them to understand that GIS is essential to their security services, and also to their day-to-day business operations. We need to show them the many kinds of value GIS delivers.

On the government side, I think we need to sharpen our arguments in favor of GIS staffing and funding and do a better job documenting benefits. I think the academic community and organizations like the NYSGIS Association are in the best position to do this. I'm quite optimistic about the future of GIS but like all innovative technologies that are not easy to explain, it will continue to be a battle to convince people of its worth. I continue to hear reports that during periods of government downsizing, it is GIS staff that often takes a hit. GIS personnel are often targeted because senior managers don't really understand all the benefits GIS provides. We must all take responsibility for fixing this.

Any specific skill sets you could recommend to somebody young in their career that they should really try to emphasize GIS or otherwise?

Well, clearly, you need to have training in geographic sciences, data management and related technical skills, mastery of spatially enabled technologies, and a strong awareness of what makes spatial data … special. Functionally, you need to be able to integrate spatially enabled data from many different sources, and turn that information into effective analytic products and visualizations that say meaningful things. GIS's greatest value comes from its ability to improve work processes and deliver essential information to business and governmental decisions and operations. To do this, GIS personnel should also have a solid understanding of business process reengineering and change management techniques. Additionally, I think that it is useful for GIS personnel to cultivate organizational and political skills. A major strength of GIS is its ability to integrate data normally kept in isolated silos, and turn it into a wide variety of valuable analytic products. To do this successfully requires that the GIS practitioner needs to understand organizational dynamics and know how to get people to work together collaboratively. If you can't find a way to convince another agency or an outside organization to share its data with you, then all the technology and technical knowledge in the world may not be enough to enable you to achieve your mission.

Antje Hecheltjen* (Chapter 4)

What advice can you give to someone new to information technology, GIS remote sensing, and international disaster management? What skills should they learn, GIS or otherwise? What advice can you give to someone who is new and who wants to get into this area?

I think programming skills are really useful. For example, JavaScript is good for web applications or for R[†] for remote sensing, if you want to get deep into it. The ability to think spatial, is essential. You also need to be able to interact with groups or individuals that have a different background and focus as yourself, since disaster and risk management is an extremely interdisciplinary field. Interact with others in an interdisciplinary way as early as possible. Try to understand who is doing what in disaster and risk management, and, of course, identify the resources and services that are available, because there is a lot out there.

---

* The views expressed herein are those of the author(s) and do not necessarily reflect the views of the United Nations.
† R Project website, http://www.r-project.org/.

How did you become a UN Junior Professional Officer (JPO)?

I was looking at job opportunities which bring my different interests together. At the university it was very technical, my tasks were basically revolving around the development of algorithms, not so much around applications. However, I had started studying geography with a focus on developing countries and on environmental issues and I was looking for a way to bring those two aspects together again. During my work at the university I had sometimes already worked as a consultant at the United Nations Convention to Combat Desertification [UN CCD]* in Bonn. So when the opportunity to work at UN-SPIDER came around, it was a perfect match. It was quite a long application process, but in the end was successful.

Michael Judex, PhD (Chapter 4)

What advice can you provide to someone new in their career that are interested in working in disaster management and GIS, either within Germany, internationally, or perhaps just in general?

Coming from the field of geography and from the perspective of GIS, I would say a major benefit is contextual thinking. What I mean is to broaden the thinking, to accept new ideas, not to have one straight line of thinking. For example, based on my training as a geographer, I see not only the environmental effects of watershed management, but I also see the socioeconomic factors. I see the psychological factors. So, it is important to incorporate all those different dimensions of human beings and the interactions of humans with the environment. This is also needed during crisis situations and for risk management. So, that would be a really important asset. Of course, the knowledge of spatial methods that are available within GIS is important, and is an advantage for the career. I think it's always good to understand data structures, spatial databases, and the heterogeneity of spatial information, which is becoming more and more complicated. Furthermore, I often observe that people coming from the university are technical oriented, so they're just looking on the technological perspective. But also very important is the user perspective. For example, if I just tell my colleagues in the situation room, "Look, I have a really nice technology, very good, you must use it" and they agree and say "Wow!, it's really nice." But they will not use it because it's not designed for their workflow. That's where the user orientation is really, really important if you want to promote new technology and in particular GIS.

I also want to mention the importance of cartography and design. I think that's a very important topic because you can do a lot of stuff with GIS and with geodata. But if it comes to the real uses, and especially during crisis situations, cartography is very important but is still very neglected because the method how you present your information has a tremendous impact on the usage. You have to figure out if the meaning of the map is recognized and used by the crisis management stuff or not. Maybe it's just too complex, too complicated, too much information on the map, or it's clearly structured and you can get the message within five seconds. So, that's always the challenge.

---

* United Nations Convention to Combat Desertification, http://www.unccd.int/.

Scott McCarty (Chapter 4)

What advice can you provide someone new in their career that is interested in working with disaster management and GIS?

It's always important to look at what you're doing and apply it to the current state. But then, I always tell the people that work here with me, you have to keep an open mind. I mean look at what the next best thing is going to be. You usually get some hints that there's going to be some newer technology coming out six months down the line or, a year down the line. When you hear about that stuff, that's the time when you start thinking about, okay, if this does come out, how can we use it and how can we apply it here with what we do. For example, what can we do to help out first responders during exercises and live events to make their job easier and to help them solve different problems that they encounter in the field? For the person that might not be into becoming a GIS analyst or technician—you don't need to be a power user to use GIS, you can use it in many different lines of work. There are many different fields out there that are finding GIS to be an important tool. That's what I kind of like because we get people calling Monroe County all the time and saying "I'm in real estate ...", which is a common one, to say they rely on our public web mapping applications for their day-to-day business. The real estate developers and the real estate agents, they use our online web mapping every day. It's someone that's probably never taken a GIS course in their life or doesn't know 99% of the GIS terminology, but they know how to run Google Maps and they know how to run our Monroe Map Viewer. So, for people that aren't coming out to be a GIS services person like myself and that are going into other fields, look to see how you could use GIS as a tool.

Jörg Szarzynski,* PhD (Chapter 4)

What advice could you provide to someone new in their career that's interested in working for the UN on disaster management and GIS? For example, somebody coming out of an undergraduate or graduate program with a degree in geography that maybe they know a little bit about GIS and they're interested in working for the UN and they're coming in with a skill set. They'd like to be a part of the UN and they are learning GIS in school. So, they'll have some skills already and they'd like to try to be a part of the UN.

In the beginning, I think this person would have to take the general decision, "Do I want to invest, at least a couple of more years, in order to become a full technical professional in GIS and Remote Sensing (RS), to cover a more technical functionality in my later position within the UN?" Such a person would be looking at those agencies doing the operational work, such as Zentrum für Sattelitengestützte Kriseninformation (ZKI) or United Nations High Commissioner for Refugees (UNOSAT) as mentioned before. If a person decides to go this path, he or she should be aware of some other time-consuming activities that are very prominent within active disaster management. For instance, you always have to be prepared to work on a 24/7 working schedule. In certain cases, such a

---

* The views expressed herein are those of the author(s) and do not necessarily reflect the views of the United Nations.

job would also take you sometimes to foreign countries where disasters take place. More or less, you would be "on call." Whenever something is happening somewhere and you would be on-call duty, you might get fielded to this area. This is certainly interesting for some, but not necessarily to all of the people. Therefore, this would be a very fundamental decision. But if you take a careful look at some of the bigger agencies, like, for example, World Food Programme (WFP), they have a known GIS department in Italy doing a lot of mapping activities carried out by a number of experts. The same can be observed within the UNHCR. They also coordinate with other institutes and universities to make sure that cutting-edge information is available whenever a disaster takes place. So, basically, a technical expert with a solid knowledge and good expertise in remote sensing and GIS can find the one other option within the UN system to bring in this expertise and to serve the numerous areas of applications that are requested in international risk and disaster management. However, the number of jobs is somehow limited. As a general rule of thumb, we always say within the UN family, we have about maybe some 500 people that are intensively working in remote sensing in GIS. That's a rather constant number over the last years. On the other hand, with the observable trend of increasing extreme weather events and the growing numbers of hurricanes and storm surges also the number of required geospatial experts might increase within the next years.

Lóránt Czárán* (Chapter 5)

What advice can you provide to someone new in their career, such as a student, that is interested in working in disaster management for GIS for the UN or perhaps just the broader international context?

You need some GIS-related courses, first. You need, definitely, some remote sensing basics, but you also need to consider courses in terms of GIS and disaster management. There might be disaster management–related course work that you also have to understand because you have to understand both sides. For example, first, what does disaster management mechanisms mean? Then, on the second hand, what the technologies can do in that respect. So, ideally, you would study on both directions, study both aspects so you understand the inter-linkages. That's important if you want to work in that domain.

If you're already an expert in GIS, I would still say you have to always look at the bigger picture, understand what different institutions, organizations are doing because you have to make sure that you're not initiating things and work and projects that are somehow duplicating or triplicating what others do, just because you don't have the full picture. You're coming into this domain, but you don't know that your sister organization on the left or other organization, UN organization on the right is doing the same work in the same country in the same context. This is, I think, key. I always see when new people come in the picture,

---

* The views expressed herein are those of the author(s) and do not necessarily reflect the views of the United Nations.

most often they don't have that institutional history, memory. They are not well prepared in terms of how the network and the "family" works. I think this happens as very few universities are actually teaching or presenting about how the UN works or what the different branches of UN or international organizations are doing. So, when you start working in any international context, you have to understand these things. You have to have some experience or you have to do your homework.

In my experiences with working with over 50 UN interns over the years, most of our interns learned everything about the organizational background and understood the dynamics between institutions and the conflicts too after some three months or so. So it is possible, doable. That's important because when you come out of such an internship, again, you have your initiation, let's say, in this domain of geospatial and disaster management. You know who's doing what, what organizations are involved, what names are counting, and where you can go to get what because that's critical. Otherwise, you can easily come in, be hired as a geospatial expert for a big NGO and then be told, go start doing support for a given country. You could then start doing things on your own without even checking if a OCHA [Office for the Coordination of Humanitarian Affairs], WFP, UNHCR, the Cartographic Section at UNHQ, or a UN Peacekeeping Mission are not doing already the same, or if they have relevant data that they can just share with you.

Then, you easily end up in situations like I've seen very often with young guys sitting in an office digitizing road networks of a country from a SPOT image.* When I would ask them, "Do you know that OpenStreetMap exists or Google Map Maker exists?" they say, "No, what is that?" You have to have your homework done in terms of having some background, understanding who is working in this area, what they are doing, where you can knock on each door, for what kind of data or what kind of support, and always have the idea of collaboration and partnerships in mind. Do not be an individualist. Don't think individualistically. Don't think selfish. Don't drive—don't run or don't go by "me being the first" or "me being the greatest," collecting all the glory. Share that "glory" if needed, but try to work together with other institutions because if your interest is to support a specific situation or a specific country or a specific community with what you're doing, ultimately that's our purpose in disaster management. Then, your first interest or primary interest should be supporting and doing the best you can for helping the people you serve rather than chasing personal glory or achievement or making your organization a little better than others and advancing your own career only.

What I'm saying is people will admire you and people will recognize you in the end, even though it might not seem so in the beginning, but in the long term they will appreciate your selflessness and your commitment to the cause rather than to the personal interest. And in the long term it pays off. I would say, too, if you're not an expert from the beginning, sure, you can become more of an expert in one or two years or at least start becoming an expert in this domain

---

* AIRBUS Defence and Space, "Spot Satellite Imagery," http://www.astrium-geo.com/en/143-spot-satellite-imagery.

GEOGRAPHIC INFORMATION SYSTEMS (GIS) FOR DISASTER MANAGEMENT

if you want, but then definitely pick up on the courses and training such as master's degrees from very specialized institutions which have already a good history of training people for these purposes. There are some good examples. It doesn't have to be in a developed country. It can be anywhere. There are good examples of such degrees which could be earned where you would be prepared for both aspects, both disaster management and GIS. Some of our Regional Support Offices (RSOs) in the UN-SPIDER network are among those training institutions offering such courses, in developing countries as well.

David Alexander (Chapter 7)
What advice can you provide someone new in their career that is interested in disaster management GIS within the US federal government?
I would say get involved with the NGOs and community groups. Don't just take classes in GIS. Try and tie GIS to the conduct of what emergency management and emergency responders do. The only way you're going to get that is if you embed. I think that's one of the things that we're probably remiss. That's one of the dangers of moving GIS to the back office. To disconnect from the operators you're really trying to support and you don't necessarily understand their rhythm, behaviors, and needs. I encourage folks to really get out there and participate.

GIS for Disaster Management Career Summary Points

The following points summarize recurring career advice ideas from the book's interviewees plus some of my own advice based on working with GIS students interested in disaster management:

- *Be interdisciplinary:* Study a wide range of topics such GIS, remote sensing, disaster management, information technology, and any other disciplines relevant to your specific disaster management interest. For example, take earth science classes for understanding earthquakes or sociology classes for understanding social vulnerability.
- *Be open-minded:* Be flexible with opportunities that present themselves. Given the diversity of GIS for disaster management, you never know when some aspect of disaster management or GIS you never considered could become a great career opportunity.
- *Get involved:* Volunteer with emergency management opportunities like FEMA or the Red Cross or technical groups like GISCorps. Disaster management is an active, vibrant field that requires real experience to become knowledgeable.
- *Stand out and go the extra distance:* More related to learning GIS, go beyond learning out-of-the-box GIS skills. Learn skills like computer programming, web development, and database management. GIS itself is very interdisciplinary. Many people take classes on how to use GIS itself but fewer learn additional IT skills that can create synergy with GIS. Learning additional IT and computing skills coupled with GIS skills will make you stand out when applying for jobs as it shows you are someone who can go the extra distance and be serious about GIS.

Staying Current in the GIS for Disaster Management Field

GIS, like any technology, is constantly changing it is important to keep up with current trends so that your skills and perspectives remain relevant. Furthermore, the disaster management field is equally in a constant state of change, as you have seen in this book, along with perspectives on new forms of data such as social media, big data, and the realities that disasters continue to grow in size, scope, complexity, and intensity. The following is a nonexhaustive list of various organizations, academic and trade publication outlets, training and education opportunities, and volunteer opportunities to consider to keep yourself current in the GIS for disaster management field. Note that many of these items are not directly related to GIS for disaster management. Instead, they are either focused on disaster or emergency management specifically and GIS is a topic that can be found with them or they are specifically focused on GIS and the topic of disaster or emergency management can be found within them.

Organizations
- International Association of Emergency Managers (IAEM): http://www.iaem.com/
- The National Emergency Management Association: http://www.nemaweb.org/
- International Network of CrisisMappers: http://crisismappers.net/
- Information Systems for Crisis Response and Management Association: http://www.iscramlive.org/
- Association of American of Geographers: http://www.aag.org/
- International Geographical Union: http://igu-online.org/
- International Cartographic Association: http://icaci.org/
- Global Spatial Data Infrastructure Association: http://www.gsdi.org/
- The Open Source Geospatial Foundation: http://www.osgeo.org/
- Open Geospatial Consortium: http://www.opengeospatial.org/

Note that many of these organizations hold conferences. You should look closely through their websites to see what they offer.

Conferences
- World Conference on Disaster Management: http://www.wcdm.org/
- International Disaster and Risk Conferences (IDRC): http://idrc.info/about/idrc-conferences/
- Geo-information for Disaster Management (Gi4DM) Conference series: http://www.gi4dm.net/
- Esri International User Conference: http://www.esri.com/events/user-conference
- GIScience Conference: http://www.giscience.org/
- Local GIS conferences: Do a Google search for GIS conferences happening in your county, state, or region; for example, from New York State: http://www.nysgis.net/activities/conferences/

Journals and Magazines
- *Directions Magazine*: http://www.directionsmag.com/
- *ArcNews* magazine: http://www.esri.com/esri-news/arcnews
- *ArcUser* magazine: www.esri.com/esri-news/arcuser

- Journal of Homeland Security and Emergency Management: http://www.degruyter.com/view/j/jhsem
- *Disasters* journal: http://onlinelibrary.wiley.com/journal/10.1111/(ISSN)1467-7717
- *Computers & Geosciences*: http://www.journals.elsevier.com/computers-and-geosciences/
- *Computers, Environment and Urban* Systems: http://www.journals.elsevier.com/computers-environment-and-urban-systems/
- *International* Journal *of Geographical Information Science*: http://www.tandfonline.com/toc/tgis20/current#.UzNtyvldWSo

Training and Education
- National Disaster Preparedness Training Center: https://ndptc.hawaii.edu/
- FEMA National Preparedness Directorate, National Training and Education: http://training.fema.gov/
- Esri Training: http://www.esri.com/training/main
- United Nations Institute for Training and Research (UNITAR): http://www.unitar.org/event/
- GIS Certification Institute: http://www.gisci.org/
- CRC Press Homeland Security book series: http://www.crcpress.com/browse/category/HSC

Volunteer Opportunities
- GISCorps: http://www.giscorps.org/
- FEMA CERT: http://www.fema.gov/community-emergency-response-teams
- American Red Cross Volunteer: http://www.redcross.org/support/volunteer
- International Federation of Red Cross and Red Crescent Societies: https://www.ifrc.org/en/what-we-do/volunteers/
- MapAction (UK-based): http://www.mapaction.org/

CHAPTER SUMMARY

In this chapter, you learned about four different things. The first part of the chapter was a nonexhaustive list of special topics related to GIS for disaster management. First, you learned about the idea of visual analytics, which is the idea of supporting analytical reasoning with interactive visual interfaces—a topic that is very relevant in disaster management in terms of supporting decision making from vast and diverse amounts of evidence. Closely related to the topic of visual analytics was a brief discussion of the idea of big data. The prime example of big data for disaster management is the incorporation of massive amounts of social media artifacts such as tweets that can be used as one potential source of situational information during disaster. You also got some ideas on specific technologies you might incorporate for working with your own big data. Next, you learned about the idea of serious games for GIS and disaster management. Serious games can be used to create scenarios and simulations, and a specific example was shown of building a serious game directly inside a commercial-grade GIS software environment in order to teach people about the capabilities of GIS and build spatial thinking skills. Finally, you were given a brief introduction to geographic information science in order to gain perspectives on the underlying science behind GISystems software, and you are encouraged to develop your own interests in geographic information science to advance the frontiers of Geographic Information Systems for disaster management.

The second half of the chapter looked to the future of GIS for disaster management based on information from several people interviewed for this book. A research agenda for GIS for disaster management was provided based on perspectives from these people and other research conducted for this book.

In the third part of this chapter, you were given extensive advice from a wide variety of people on developing a GIS for disaster management career and how to find a job in the field. In particular, make note that many of the people recommended an interdisciplinary approach to pursuing a job in the GIS for disaster management field. Thus, it will be important to keep an open mind and be flexible in pursuing your career.

The final part of the chapter gave you a list of organizations, conferences, journals and magazines, training and education, and volunteer opportunities that you can follow up on to keep yourself current in the GIS for disaster management field.

I hope you have enjoyed reading and learning from this book. Ideally, this book has given you ideas, inspiration, education, and training to think spatially and start or advance your career in the diverse, vibrant, and exciting field that is GIS for disaster management.

DISCUSSION QUESTIONS AND ACTIVITIES

1. Do some Internet searching and see if you can find some other examples of a visual analytics system designed to work with big data. What types of visual interfaces does the system use? If the system is not directly for disaster management, how might you adopt the system to support disaster management activities like those that you learned about in this book?

2. If you were to design a serious game for GIS and disaster management for people new to GIS, what type of scenario might you create and what types of GIS actions would you include in the game to match your scenario? Think about what you've learned this book and your own experiences with GIS.

3. Based on what you read in this book, and your own research, what do you think the future of GIS for disaster management will look like and why?

4. If you are looking for a job in the GIS for disaster management field, review your resume and see if your resume and matches some of the things that the people who provided advice on finding a job in the GIS for disaster management industry recommended. What, if any, potential gaps do you see in your experiences, and how might you fill those gaps if necessary?

5. As discussed in the section on staying current in the GIS for disaster management field, many of the items listed are not directly related to both GIS and disaster management. Take a look at some of those items, and see where you can find GIS imbedded inside of a disaster management item, and the do the reverse of finding a disaster management item imbedded inside of a GIS item.

RESOURCES NOTES

For a video overview of the visual analytics system shown in Figure 9.1 and the research it was developed for, see https://www.youtube.com/watch?v=mo3fLuWHQnM.

For more discussion on the role of social media in disasters, see A. Crowe, A. 2012. *Disasters 2.0: The Application of Social Media Systems for Modern Emergency Management.* Boca Raton, FL: CRC Press.

For an example of using crowdsourcing to identify items of interest from imagery, see the tomnod website, http://www.tomnod.com/nod/.

For an example of a cutting-edge United Nations organization focused on the use of big data for development in helping vulnerable people, see the United Nations Global Pulse program, http://www.unglobalpulse.org/.

For Esri perspectives on big data, see http://www.esri.com/products/technology-topics/big-data.

For a thorough guide to noSQL databases, see http://nosql-database.org/.

For some examples of games that can be used teach disaster management concepts (but not necessarily with a direct GIS component) see The Stop Disasters! Game from UNISDR, http://www.stopdisastersgame.org/.

REFERENCES

Ahlqvist, Ola, Thomas Loffing, Jay Ramanathan, and Austin Kocher. 2012. "Geospatial human-environment simulation through integration of massive multiplayer online games and geographic information systems." *Transactions in GIS* 16 (3):331–350.

Blochel, Kevin, Amanda Geniviva, Zachary Miller, Matthew Nadareski, Alexa Dengos, Emily Feeney, Alyssa Mathews, Jonathan Nelson, Jonathan Uihlein, Michael Floeser, Jörg Szarzynski, and Brian Tomaszewski. 2013. "A serious game for measuring disaster response spatial thinking." *ArcUser* 16 (3):12–15.

Cheng, Zhang, Fang Hao, Zeng JianYou, and Shuai Yun. 2010. "Research on design of serious game based on GIS." Paper read at IEEE 11th International Conference on Computer-Aided Industrial Design and Conceptual Design (CAIDCD), November 17–19, 2010.

Cox, Richard. 1996. "Analytical reasoning with multiple external representations," PhD thesis, Department of Artificial Intelligence, University of Edinburgh, ftp://crivvensvm.inf.ed.ac.uk/pub/graphics/rcox_thesis.pdf (accessed June 9, 2014).

Cox, Richard. 1999. "Representation construction, externalised cognition and individual differences." *Learning and Instruction* 9 (4):343–363.

Crowley, John. 2013. "Connecting grassroots and government for disaster response," Commons Lab, Wilson Center, http://www.wilsoncenter.org/publication/connecting-grassroots-and-government-for-disaster-response-1 (accessed June 9, 2014).

Dibiase, D., M. DeMers, A. Johnson, K. Kemp, A.T. Luck, B. Plewe, and E. Wentz. 2006. *Geographic Information Science and Technology Body of Knowledge.* Washington, DC: Association of American Geographers.

Fuhrmann, Sven, and William Pike. 2004. "User-centered design of collaborative geovisualization tools." In *Exploring Geovisualization*, edited by J. Dykes, A. MacEachren and M. J. Kraak. New York: Elsevier.

Gartner Inc. 2011. "Gartner says solving 'big data' challenge involves more than just managing volumes of data," Gartner Inc., http://www.gartner.com/newsroom/id/1731916 (accessed March 26, 2014).

Jacobs, Adam. 2009. "The pathologies of big data." *Communications of the ACM* 52 (8):36–44.

Joint Board of Geospatial Information Societies (JB GIS). 2013. *The Value of Geoinformation for Disaster and Risk Management (VALID): Benefit Analysis and Stakeholder Assessment*, United Nations Office for Outer Space Affairs Platform for Space-Based Information for Disaster Management and Emergency Response (UN-SPIDER), http://www.un-spider.org/VALID-stakeholder-assessment-I (accessed June 9, 2014).

Larkin, Jill H., and Herbert A. Simon. 1987. "Why a diagram is (sometimes) worth ten thousand words." *Cognitive Science* 11 (1):65–100.

Mark, David M. 2003. "Geographic information science: Defining the field." In *Foundations of Geographic Information Science*. New York: Taylor & Francis, 3–18.

Michael, David R., and Sandra L. Chen. 2005. *Serious Games: Games That Educate, Train, and Inform.* Boston, MA: Thomson Course Technology PTR.

mongoDB. 2014. "Map-reduce," mongoDB, http://docs.mongodb.org/manual/core/map-reduce/ (accessed March 26, 2014).

mongoDB. n.d. "NoSQL databases explained," mongoDB, http://www.mongodb.com/learn/nosql (accessed March 26, 2014).

Robinson, Anthony C., Jin Chen, Eugene J. Lengerich, Hans G. Meyer, and Alan M. MacEachren. 2005. "Combining usability techniques to design geovisualization tools for epidemiology." *Cartography and Geographic Information Science* 32 (4):243–255.

Thomas, James J., and Kristin A. Cook. 2005. *Illuminating the Path: The Research and Development Agenda for Visual Analytics.* Los Alamitos, CA: IEEE.

Tomaszewski, Brian, and Alan MacEachren. 2012. "Geovisual analytics to support crisis management: Information foraging for geo-historical context." *Information Visualization* 11 (4):339–359.

US Department of Health and Human Services. 2014. "What & why of usability," Usability.gov, http://www.usability.gov/what-and-why/index.html?utm_source=twitterfeed&utm_medium=twitter (accessed April 6, 2014).

Zhang, Jiajie. 1997. "The nature of external representations in problem solving." *Cognitive Science* 21 (2):179–217.

Zhang, Jiajie, and Donald A. Norman. 1994. "Representations in distributed cognitive tasks." *Cognitive Science* 18 (1):87–122.

INDEX